Grasberg

Grasberg

MINING THE RICHEST
AND MOST REMOTE DEPOSIT
OF COPPER AND GOLD
IN THE WORLD, IN THE
MOUNTAINS OF IRIAN JAYA,
INDONESIA

George A. Mealey

Produced and published by
Freeport-McMoRan Copper & Gold Inc.
1615 Poydras Street
New Orleans, LA 70112

Copyright © 1996 Freeport-McMoRan Copper & Gold Inc.
ALL RIGHTS RESERVED
Printed in Singapore
ISBN: 0-9652890-0-1

Editor: David Pickell

Research: Kal Muller
Book design: David Pickell
Jacket design: Peter Ivey
Cartography and graphics: David Pickell
Photography: Kal Muller and others as credited
Production liason: Mary Chia, Periplus (S) Pte. Ltd., Singapore

Set in Quadraat
Color separation by PICA Colour Separation (Pte.) Ltd., Singapore
Printed and bound by Tien Wah Press (Pte.) Ltd., Singapore

To Jean
*who over the years has put up with me
and life with a mining engineer*

Contents

LIST OF MAPS, FIGURES, AND TABLES x

PREFACE xi

Part I
DISCOVERY AND RISK

1. Freeport and Irian Jaya

A mountain of copper and gold—Herman Frasche—The Freeport Sulphur Company—Cuban nickel—A history of Indonesia—The Suharto era—An unknown island—The Dutch—Irian Jaya—Modernization 21

2. The Discovery of Ertsberg

A shocking visit—Terra incognita—British ornithologists—The race to the snow—Oil exploration—The Colijn expedition—Ertsberg—A forgotten report—Forbes Wilson—Gleaming chalcopyrite—13 acres of ore 51

3. Building a Mine

Turbine helicopters—Irian Barat—New leadership in Indonesia—A contract of work—Drilling Ertsberg—Darnell Ridge—Boardroom doubts—An impossible road—The first concentrate shipment—Tembagapura 79

Part II
CRISIS AND EXPANSION

4. The Underground Era

Marginal profitability—Steepening the pit—Vexatious deposits—Ertsberg East—The Filipino miners—A stubborn block cave—The ore passes—Gold fever in the boardroom—A spectacular core sample 109

5. The Grasberg Era

A mistaken assumption—The porphyry model—A disappointing test hole—Gold—Rapid expansion—Conveyors and ore passes—The black chicken—An ancient volcano—Block A riches—Exploration challenges 137

6. To 115,000 Tons a Day

The Heavy Equipment Access Trail—The 115K program—Leveling a ridge—The SAG mill—A road through a swamp—Raising $1 billion—The privatization program—Creativity and risk—Record production levels 165

Part III
THE BUSINESS OF MINING

7. Mining the Ore

Shrouded in fog—Four billion tons of rock—Poker chips—Running the 'Pay-Hah' 4100—The crusher—Underground mill sweetener—Block cave mining—The 'rocks in the box'—Concentrators—The slurry line 189

8. Exploration

Plate tectonics—A plumbing system—The New Guinea mineral trend—Aeromag—Hoisting—Wabu Ridge and the 'Fighting Irish'—Etna Bay—Jurassic Park—Life in the field—Some ugly rock—A new drill rig 215

9. The Use and Sale of Copper

An ancient metal—The Cyprus copper works—Copper and electricity—The flotation revolution—Smelting and refining—Heap leaching—A recyclable metal—The custom market—World demand 247

Part IV
RESOURCES AND THE FUTURE

10. Managing the Environment

Muddy rivers—Sheet flow—Tailings deposition—Overburden—The shrinking glaciers—A biologically rich setting—hydro-mulching—Farming the tailings—Recycling—Environmental monitoring 265

11. Community Relations

A special obligation—The Kamoro—Missionaries and the Dutch—The Amungme—Cargo cults—Land ownership in Irian—Health care—The business incubators—Expectations—Sustainable development 290

12. Timika and a New Town

The empty jungle—An airport—Spontaneous settlements—Kwamki—A booming area—Transmigration—Entrepreneurial spirit—Kuala Kencana—A model Indonesian town 321

13. The Future

Bright prospects for growth—Exporting technology—A bridgehead—Skilled local work force—Expanded provincial infrastructure—An agricultural center—Continuing demand for copper 340

APPENDIX 355

ACKNOWLEDGEMENTS 372

INDEX 373

Maps, Figures, and Tables

Fold-out maps
Project Area 1:500,000 .... *after page* 96
Mine Area 1:62,500 .... *after page* 144
Map 1.1 Indonesia 1:50M 33
Map 1.2 Irian Jaya 1:8M 41
Map 2.1 Exploration 1:1.2M 53
Map 5.1 Mine Area Geology 1:60,000 152–153
Map 6.1 Tembagapura ~1:10,000 180
Map 7.1 The Mill ~1:3,500 208–209
Map 8.1 New Guinea Tectonics 1:20M 217
Map 8.2 Mine Area Faults 1:125,000 225
Map 10.1 Tailings Area 1:200,000 269
Map 10.2 Carstensz Glaciers 1:50,000 277
Map 11.1 Banti Village ~1:6,500 309
Map 12.1 Timika Town ~1:15,000 328–329
Map 12.2 Kuala Kencana ~1:20,000 333
Figure 4.1 Mine Production 1973–1994 117
Figure 4.2 Mill Rate 1973–1995 121
Figure 4.3 Schematic Illustration of Block Caving ... 125
Figure 5.1 Formation of the Grasberg Complex 157
Figure 7.1 Plan View of Mine Operations 1:30,000 193
Figure 7.2 Cross-Section of Mine Operations ... 200–201
Figure 9.1 Price of Copper 1969–1995 248
Figure 9.2 Price of Gold 1969–1995 252
Figure 9.3 Price of Silver 1969–1995 253
Figure 9.4 Copper Consumption 1995 259
Figure 9.5 Treatment and Refining Charges 1982–1995 ... 260
Figure A-1 Schematic of Ore Flow 356
Figure A-2 Final Grasberg Pit *with overburden stockpiles* 360
Figure A-3 Ore Passes 361
Figure A-4 North and South Concentrator Circuit 364
Figure A-5 SAG Mill Circuit 365
Table 5.1 Grasberg Drill Hole Timing and Results 148
Table 5.2 Proved and Probable Reserves 1995 162
Table A-1 General Reagents Usage 371

Preface

I CAN REMEMBER quite clearly the first time I heard of the Ertsberg mining district. I was working in South America in a Chilean copper mine called El Teniente. As far as I was concerned, El Teniente, along with Chile's other large-volume, high-production mines, was where the action was in the field of copper mining, and I considered myself a copper miner. The world's largest, richest and most productive mines were dominated by five companies: Kennecott, Anaconda, ASARCO, Phelps Dodge, and Newmont. All of these companies were active in Latin America, and the Chilean mines set the pace for the industry. My standard for measuring the quality of any operation in the world was set by the Chilean mines.

The year was 1968 and then—as it still does today—Chile provided much of the world's copper. But I had no way then of knowing that the world of copper would soon turn upside down. The fortunes of the major companies would change drastically, and one company would disappear. Nor could I know that a ridiculously small deposit I had noticed in the technical press would not only take on a great importance in my own life, but would become a producer surpassing anything seen in the 1960s, and by the 1990s become one of the top three mines in the world.

As I look at the Ertsberg mining district today, I have every reason to believe the operation has not topped out. There are plenty of resources and opportunities for continued growth, and I look forward to watching the next 20 to 30 years.

My first notice of the Ertsberg district was in *Mining Engineering*, a technical publication for the mining industry. The journal mentioned that the Freeport Sulphur Company had just completed the diamond drilling of a deposit in the interior of New Guinea, having encountered reserves of 30 million tons of 2.5 percent copper. The company had completed a feasibility study, and intended to construct production facilities designed for 6,500 tons per day. The project cost, including infrastructure, would be in excess of $100 million.

To me a project this expensive, this small, and in this location appeared to be sheer madness. Although the grade was a little higher than what I was accustomed to seeing in South America, I could not believe that anyone would undertake such a project. I was then working as a project manager for a 25,000-tons-per-day expansion of the El Teniente facilities. Our ore grade was 1.85 percent copper, and the project cost, including all infrastructure, came to $185 million. Similar projects were underway elsewhere in Chile, at Andina and Chuquicamata, and in Peru, at Cuajone. Even correcting for their slightly better grade of ore, Freeport's costs could be more than 150 percent of ours in South America—and this for a new, unproven operation. I was convinced that either the people at Freeport Sulphur did not know what they were doing, or there was some sort of stock price manipulation scheme going on. I made a mental note to follow this project to see what kind of a disaster resulted.

As time passed, my fortunes and outlook for the future in Chile began to change. By 1970 the project I was overseeing was drawing to a close. It was shaping up to be a great success, but the political situation in Chile was fast deteriorating. The leftist candidate Dr. Salvador Allende Gossens won election to the presidency, and his socialist Popular Unity government planned to nationalize the Gran Mineria—the large, foreign-owned copper mines. Regardless of how much I loved Chile, it was clear there was no future there for gringo George Mealey. It was time for me and my family to get out, and we hurriedly returned to the United States.

In late 1970 I found myself on the job market—together with hundreds of other mining engineers leaving Chile. The economy in the United States was not good. For a while I found employment with Fluor-Utah, a California-based engineering and construction contractor that was active in the mining field. Business was not good for the company at that time, and I became discouraged. Even though I enjoyed my job at Fluor-Utah, I decided to find a position working directly for a mining company, and took a job with Denison Mines in Ontario, Canada. Denison was a leader in the field, and their large-scale underground operation at Elliot Lake was a major producer of the world's uranium. This was a wonderful opportunity to broaden my experience.

While in Canada I continued to follow the fortunes and follies of the Ertsberg project with keen fascination. I read in the technical press of the great difficulties encountered in construction. I noted the large cost overruns. But I also noted that they kept to their schedule, and they actually got into production by late 1972, like they said they would. I had to give the Freeporters credit for getting a job done. They actually began to ship copper. Still, I was sure the mine could not be profitable, and that the enterprise would fail.

Then a remarkable event occurred. The year was 1973. The Vietnam War was at its peak, and Richard Nixon was in office. Nixon instituted domestic price controls that pegged the price of copper at 70 cents a pound. The effect of this was to make the world market, as measured by London Metal Exchange (LME) prices, go crazy. The Vietnam-era demand for copper drove the LME price—which was outside Nixon's sphere of influence—to $1.40 per pound. Under these conditions, the Ertsberg operation, which sold its copper outside the United States based on LME prices, began to prosper. In my wisdom I knew that copper prices could not stay this high, and in the end I was sure my initial judgment was correct—the operation was too costly and would fail.

By the middle of 1973 my family and I decided to return to the United States, as I had a desire to obtain an MBA in management. I was able to work out an arrangement with Fluor-Utah in San Mateo, California, where I could work full-time as an engineering manager and, if I had the physical endurance, take an evening and weekend program in business administration in the San Francisco Bay Area.

While working with Fluor I continued to follow developments at the Ertsberg mine. Freeport's prime contractor was Bechtel, and they were one of Fluor's toughest competitors. Fluor, in fact, was eager to take on the Ertsberg project. I discovered a report in the company files that the Freeport Sulphur Company had asked Fluor to produce as a second opinion of Bechtel's recommendations and cost estimates. (I learned later that this study almost won Fluor the entire construction job from Bechtel.) In conversations with several of the Fluor engineers who had worked on the report, I learned that in their opinion the potential of the Ertsberg mining area was tremendous. They were, in fact, privately somewhat critical of Freeport Sulphur for not being more aggressive in its exploration program, and for not bringing expanded operating facilities on line. They had tried to sell this strategy to Freeport, but had failed.

By mid-1975 I graduated with my MBA and things again began to change in my life. I was approached by Milt Ward, a vice president of Freeport Minerals (formerly Freeport Sulphur), who was looking for somebody to head up a new mine development unit. This was to be a small group of minerals-oriented engineers who would be charged with evaluating new mineral deposits, bringing new deposits into production, and solving problems with existing operations. I was afraid Milt's real objective was to get me involved with Ertsberg, so I was a bit wary of his offer. I still was not convinced that Ertsberg was a project I wanted to have anything to do with.

Milt came out to California to interview me, and checked out my credentials through mutual acquaintances in the industry. Finally he asked me to come to New York for a more formal interview. He assured me that should I take the job, the scope of my activities would be very broad. My work would include problem-solving at Ertsberg, but primarily I would be responsible for bringing on new operations. He offered a very good salary, and the potential, if I was successful and if the company grew in the minerals field, for some fast-track moves into senior management. I had an agonizing decision to make. Should I take the job with Freeport? Or should I stay with Fluor? I was very happy working for Fluor—the opportunities were good, and I was progressively being given more responsibility. But after an agonizing weekend, I decided to take the Freeport job. My family supported me in this move completely.

The new position required that I move to Louisiana, which to a mining engineer in the field of metals is very far from the mainstream of activity. September of 1975 found myself and my family in New Orleans, but within two weeks I was on my way to Indonesia. There, wearing my new manager of mine development hat, I would be evaluating the potential of mining the roots of the Ertsberg deposit from underground. I must admit to considerable apprehension, as from this first trip it looked like the job was going to involve a lot more of Ertsberg than I had counted on. On the other hand, the operation had fascinated me for so long that I was truly excited about the opportunity to finally have a look at it.

IN RETROSPECT, I count it a blessing that I made the trip. As each day passed, my interest in the Freeport operation grew. The potential here indeed was tremendous. I saw a capacity for great growth—in the mine operation, in the ore reserves, and in my own responsibilities as well. All of this has turned out to be the case, and it seemed to happen faster than anyone could have imagined.

Change has been the one constant factor during the twenty years that I have worked with Freeport Indonesia. And the rapidity of this change has been staggering. For many years it has been necessary for me to travel from New Orleans to Irian Jaya six or eight times a year, and despite the frequency of my visits, at every arrival I have been impressed with the changes I have seen. This growth has occurred not only in the company's mining operations, but also in the surrounding infrastructure and community.

Where once there was a storage yard for materials, now there is a water tank. Where once there was a single-family house, now there is a four-story apartment building housing sixteen families. Where once there were indigenous people living in poverty, now there are the same people, neatly clothed and living in solid houses they built themselves, either employed by our company or earning wages in one of the many cottage industries that have sprung up around Timika.

The health, education and general living standards in the corridor around the Freeport operations have moved into the modern age. Our area has become notorious throughout the province as a place where there is opportunity and a better life. As a result, the growth of the community around us—in ways not directly related to the mine operations—has been phenomenal. Some years back a visiting Indonesian government minister called our mining operation "the honey that draws the ants." This is an accurate characterization, and our project zone has become one of the fastest growing areas of Indonesia outside of Jakarta.

When a Dutch geologist first noticed Ertsberg in 1936, and even when mine construction began in the late 1960s, no more than a few hundred people lived in the area. Today our project zone has a population of 50,000–60,000 people.

Perhaps because these changes have occurred so rapidly, I have observed that the collective memory of people working in our operation, as well as that of those living in the surrounding community, seems to date back no more than a few years. The historical knowledge of newcomers begins with their arrival, and it is as if the entire structure and social fabric of the place began when they got here.

I have found that the mining industry is full of graduates of our "school of mines"—or "school of hard knocks"—in Irian Jaya. These alumni, both Indonesians and expatriates, continue to closely follow the progress of our operation. As I travel around to industry events, I am constantly meeting these old friends who have worked for us in the past, and who have contributed much to the company. During these encounters, I am always questioned about what has transpired at the mine since they left.

It is partly my recognition of the need to keep our old friends informed that has inspired me to think about the events of my 20 years with Freeport, and to collect these thoughts in this book. The book is also the product of a desire to create a record of what has been accomplished, for the benefit of those who are yet to come. Otherwise, I am certain that the achievements of some of the finest people I have met would have been forgotten.

The alumni of P.T. Freeport Indonesia are many. There is not space here to mention them all, or the very major contributions they have made to the enterprise over the years. However I would like to make special mention of a few individuals with whom I worked closely during my time with PTFI.

Ali Budiarjo was the president of PTFI during my first ten years with the company, and he continues as an honorary director. Ali was a wonderful instructor and guide, for me and all the other expatriates at PTFI. He has that rare ability to operate in both the western and eastern worlds. Of all the Indonesians I know, he is one of the few who can say "no" in a way a westerner can understand.

Milt Ward was my direct supervisor while both of us moved up through the organization during the 16 years he was with the company. Milt brought a rare combination of interests and abilities to the company, and we were all lucky to have him at the helm. Problem-solving was just one of his many talents, but perhaps his record speaks most eloquently for him: when Milt arrived on the scene PTFI operated a small mine in danger of closing; when he left, as president and chief operations officer of Freeport-McMoRan, PTFI operated a healthy, world-class mine.

Les Acton served two full tours of duty as general manager of the Irian Jaya operations. In its day, this was probably the toughest, most stressful job in the entire Freeport family of companies. It was under his guidance that the operation experienced its first major expansion, from a small open pit mine to a medium-sized underground block-cave mine. This involved nearly tripling production, from 7,500 tons of ore per day, to 20,000 tons per day. To our great satisfaction, the underground mine—an exceedingly stubborn operation—was brought on with a minimum of accidents. The groundwork he laid in the realm of mine safety can be seen in PTFI's current safety record, which is one of the best in the world.

Tommy Williams is another general manager who devoted nearly ten years to the operation. He thoroughly understood every facet of the operation and was a tireless worker. He was in charge of the mine during a period when the company faced an extended run of low copper prices, and severe measures were required to keep the operation alive. It was also on his watch that Grasberg was discovered, setting into motion a series of changes that rapidly expanded the operation and shifted the company's outlook from short-term survival to a program of growth lasting deep into the next century.

Bob Russell is another former general manager, and a man I have known since my college days. Bob's responsibility was to oversee an ongoing construction and design engineering program that took us in stages from 20,000 tons per day to 125,000 tons per day. During the five years it took to achieve our goal, Freeport was running probably the largest and most difficult construction program in all of Southeast Asia. Bob's work took place both in Irian Jaya and at corporate headquarters in New Orleans, and he excelled in both arenas.

Jim Bob Moffett, chief executive officer of Freeport-McMoRan, was one of the first to grasp the vast potential lurking in Freeport's Indonesia asset. His knowledge of basic geology served him well in this regard. As this potential began to unfold, he quickly took charge of negotiations and discussions with the Indonesian government, both extending our contract of work and obtaining additional acreage for minerals exploration. He also recognized that the structure and corporate culture of PTFI, built upon a short-term outlook, needed to change to something more suited to long-term success. In his own way, Jim Bob has always been my strongest supporter.

Hoediatmo (Dick) Hoed recently retired as president of P.T. Freeport Indonesia after having seen the enterprise through five years of its greatest growth and change. He joined Ali Budiarjo's law practice as a young law school graduate just before Freeport came to Indonesia, and since then his career has grown with PTFI. Dick helped draft the original contract agreements between Freeport and the Indonesian government, and played a role in every agreement and most of the contracts since. When it came to relations with the Indonesian government, he was our man in the trenches, and he got the job done. Through the years, as Dick and I both moved through various jobs and responsibilities, he always provided me with wise counsel that extended far beyond the field of law.

Usman Pamuntjak was among the first ten Indonesians hired by Freeport in anticipation of the beginning of operations. At the time he was a young engineer with some experience in coal mining, but only a textbook understanding of copper mining. His first assignments were in construction, and he was there for the painful startup of mining operations using castoff construction equipment. He saw the Ertsberg begin to disappear as a junior mining engineer, and saw the discovery of Grasberg as president of PTFI. His career should serve as an inspiration to all young engineers. I met him during my first trip to Indonesia, and have always been impressed by his drive to learn and succeed in the corporate environment. Over the years he gave me excellent guidance as to how things were done in Indonesia.

Forbes Wilson and Jack Hall deserve special mention. The Irian Jaya Mine would never have existed had it not been for these two men. When I joined the company, Forbes had left full-time employment with Freeport, but remained as a director—and for me, a mentor. Much to my delight, I discovered that his first experience in mining was also at El Teniente, and that we had shared many of the same experiences there. Jack Hall also provided direction and encouragement in my early years at Freeport. He came to the company when the construction was just beginning in Indonesia, and as executive vice president of Freeport Minerals, guided the operation through the early startup period. In the beginning, Jack was the only person in Freeport who had any significant experience in the copper business. He was the man who made the hard decisions in the early days, and is responsible for PTFI's solid foundation of operating principals. Both of these men are now gone, and much missed.

The list goes on, but there are too many to name. The stage is set for the next century. An undertaking of this magnitude can be accomplished only with a cast of thousands.

PART I
Discovery and Risk

Freeport's Irian Jaya operations. At top left is Grasberg; at lower right are the mill-works; in the gorge above the mill is the old Ertsberg pit.

CHAPTER ONE

Freeport and Irian Jaya

A mountain of copper and gold—Herman Frasche—The Freeport Sulphur Company—Cuban nickel—A history of Indonesia—The Suharto era—An unknown island—The Dutch—Irian Jaya—Modernization

IN THE SHADOW of abrupt limestone ridges rising more than a thousand meters above the tropical vegetation of southwest New Guinea lies an ore deposit with a future sales value of $77 billion. This mountain of copper and gold ore stood for three million years while nearby glaciers, advancing and retreating according to the fluctuations of the world's climate cycles, scoured deep valleys around it.

An occasional group of Papuans, as the indigenous people of New Guinea are known, may have stopped to look at this mountain as they crossed the central range to trade, or as they wandered the high valleys in search of game. Although it was gently rounded and several hundred meters lower than the highest of the surrounding peaks, the mountain would have stood out because it was strangely barren of trees. Although they probably noticed this geological feature, the cold would have prevented the Papuans, who wore only penis gourds, from lingering at such altitudes.

In 1936, on a trip to scale the glaciers a few kilometers to the east, a young Dutch geologist named Jean Jacques Dozy noted the mountain on his map. He recognized its anomalous vegetation, and dubbed it Grasberg, "Grass Mountain" in Dutch. Although Dozy and the other members of the small expedition were exploring on their own time, with the purpose of being the first to scale the glaciated Carstensz mountains, as a geologist he couldn't help taking notes and collecting a small number of rock samples.

The samples were distributed to labs to analyze, and the text of this analysis, together with Dozy's own description of the general stratigraphy of the region, was published in the Leiden Journal of Geology (*Leidsche Geologische Mededeelingen*) in 1939. With the coming of World War II, the few copies of this journal that had been printed languished, unnoticed, in university libraries. Although Grasberg rated a mention in this volume, it was another, more dramatic, anomaly that attracted the most attention: Ertsberg, "Ore Mountain," a glistening black outcrop of pure high-grade copper ore standing 180 meters above an alpine swamp.

Ertsberg, probably the largest above-ground ore deposit in the world, was the magnet that drew Freeport to Irian Jaya. Freeport Sulphur exploration manager Forbes Wilson saw Dozy's report on a trip to Europe in the late 1950s, and was so impressed that in 1960 he mounted an expedition to see this wonder for himself. Wilson found Ertsberg to be every bit as impressive as the report suggested, and he convinced the Freeport board to go ahead and mine the deposit. After a major change in Indonesia's political leadership, and after overcoming countless engineering challenges, Ertsberg began producing in 1972. Grasberg, though always looming just across the broad alpine swamp, was in those days nothing more than part of the scenery.

Ten years after mining had begun, Freeport's Irian Jaya operations were in trouble. In retrospect this seems strange, as even though Ertsberg was nearly exhausted, some very rich nearby deposits had been discovered that were being mined from underground. Many of the company's managers believed there was even more copper in the area; the geological indications were strong.

The problem was that copper prices in the early 1980s were very low, hovering just above the cost of extraction. Perhaps even more important, this was a time when fiber optic cable and aluminum wire were being touted as the replacement for copper. Investors saw little future in copper mining operations, and some in the Freeport boardroom thought the best strategy would be to sell the Irian Jaya operations, and concentrate on more profitable enterprises elsewhere.

But the mood at the company began to change in 1986 when James Robert "Jim Bob" Moffett, the newly appointed chief executive officer, gave orders throughout the Freeport family of companies to renew exploration efforts. For years, geologist Dave Potter and others had been looking inquisitively at Grasberg from the makeshift geology office at the edge of the Ertsberg mine area. Potter had a strong hunch about this homely mountain, which was shared by myself and some others at the company. With no budget for a proper drill program, in 1985 he dragged an old rig over to the base of the mountain and drilled a 200-meter test hole into its side. The results from the cores were discouraging, but this did not dissuade Potter, and in 1987 he found time to take a helicopter to the top of the mountain to collect some surface samples. Subsequent assays showed these to have unusually high gold values.

By 1988, the combination of Jim Bob Moffett's exploration-oriented thinking and the rising interest of investors in gold prospects were enough to justify a five-hole program of drilling Grasberg properly, from the top. The first four holes showed good values for both copper and gold, but near-surface gold enrichment was not present, as we had hoped. The last hole, however, was a stunner: of its 611-meter length, 591 meters intercepted ore, with an average of 1.69 percent copper, and 1.77 grams per ton of gold. This may be the most impressive core ever produced in the mining industry.

With these results, everything changed for us in Irian Jaya. We moved from an operation struggling to stay alive to one in rapid expansion. Instead of the operator of a small, technically interesting, but dwindling mine in the highlands of Irian Jaya, Freeport became the custodian of the third-largest copper reserves and the single largest gold reserves of any mine in the world—as of late 1995, 40.3 billion pounds of copper, and 52.1 million ounces of gold. Our production capability has expanded such that in early 1996 we now produce, on a daily basis, more tons of copper concentrate than we mined raw ore in the first year of the Ertsberg mine. Our mill is state-of-the-art, and we are perhaps the lowest-cost operator in the world. The latest estimates show that Freeport will continue to be a profitable mine in Irian Jaya for at least 45 years.

TODAY'S FREEPORT-MCMORAN Copper & Gold, and its subsidiary P.T. Freeport Indonesia, began eighty years ago with a sulfur company operating along the Gulf Coast of the United States. Sulfur has been in use for more than 2,000 years, but it was only with the industrial revolution and the birth of the chemical industry in the 18th century that mining "the stone that burns" began in earnest. Sulfur's status as one of the building blocks of the industrial age led to the formation of multinational mining companies, among them Freeport Sulphur.

World demand for sulfur escalated dramatically in the 19th century, but the only source of this material was native sulfur from volca-

The Bryanmound operation, which seems almost quaint today, was the first sulfur mine in Texas and the beginning of the Freeport company. Photograph circa 1920.

noes. Throughout the last century, 95 percent of the world's supply was provided by a cottage industry of small mines working the deposits southwest of Mt. Etna on the island of Sicily. By today's standards, the conditions of work would be considered intolerable, with the widespread use of child labor and exposure to noxious fumes.

Then, on Christmas Eve of 1894 in the state of Louisiana, a German immigrant named Herman Frasch completed an experiment which was to change the sulfur industry forever. After an initial career in the oil industry, where he had been recruited by John D. Rockefeller to be the first director of research for Standard Oil, Frasch became interested in the possibility of mining subterranean sulfur by injecting superheated water into the deposits to melt the sulfur, and then pumping the molten sulfur to the surface.

Frasch had obtained the mineral rights to a long-known sulfur deposit over the top of a salt dome in Calcasieu Parish in southwestern Louisiana. With the success of his experiment, his Union Sulphur Company soon became profitable. Within another 20 years, the Frasch process, as carried on by others, became so successful that the

Loading sulfur at Hoskins Mound, in the late 1930s. The molten material was solidified in large bins, and blasted loose with dynamite before shipment.

center of world sulfur production shifted from Sicily to the Gulf Coast of the United States.

Although Frasch was a visionary who developed a process that rebuilt the industry, as an entrepreneur he was somewhat short-sighted. His Union Sulphur Company was highly profitable, but the Calcasieu deposit was his only asset, and it was quickly being depleted. Oil explorers in Texas discovered sulfur at a location called Bryanmound, and the owners of this property had for several years tried to sell it to Frasch. He knew his extraction process was protected by patents, and as he already considered himself to own the "best" deposit, he did not see the need to expand his resources.

He let Bryanmound and other new mine opportunities pass, which effectively sealed the fate of his corporation. In the end his enterprise failed, because any natural resource company that expects long-term life and growth must continually renew its resource base or cease to exist. Other entrepreneurs saw additional sulfur potential in the salt domes along the Gulf Coast. In 1911, a second-generation Texan by the name of Eric Swenson created a syndicate which, on July

12, 1912, resulted in the formation of the Freeport Sulphur Company.

Freeport's charter was to construct the United States' second Frasch process sulfur mine. This was, for that time, a gigantic undertaking. It involved the construction of a "free port" at the mouth of the Brazos River, on the Gulf of Mexico due south of Houston. This city was to rival Galveston and Corpus Christi as an ocean outlet. A town site was laid out for company employees, complete with a large hotel called the Tarpon Lodge—not the last hotel Freeport would build—and a three-mile canal was excavated to supply water to the town and the sulfur mining operation. The agreement with the property owners required that Freeport Sulphur complete all construction by June 1, 1913. With a remarkable mobilization of construction resources, goods and materials, the mine reached production far ahead of schedule, and the first sulfur was pumped on November 12, 1912. With this beginning, Freeport entered the mining industry as one of the major producers of the world's sulfur, a position the company maintains to the present day.

Freeport at the outset saw value in expansion, growth, and in maintaining an inventory of reserves. At first the company focused on sulfur, and by 1923 a second property came into production at Hoskins Mound, Texas. This deposit, containing nearly 11 million tons of sulfur, was almost twice the size of Bryanmound. Hoskins Mound was not an easy mine. At this site Freeport became among the first to inject mud into the sulfur formation to prevent the escape of mining water through cavities and channels, and became the first to perfect the use of sea water in Frasche process mining.

In 1932, Freeport's search for new reserves led the company to Louisiana. The Grande Ecaille property there would become the longest-lived producing sulfur mine in the state, and the second-largest sulfur mine in the world. Anticipating the depletion of Bryanmound, Freeport bought the sulfur rights to Grande Ecaille from three petroleum companies. The company quickly drilled to establish the reserves, and after rounding up a source for construction capital, began to develop the dome.

During this era, Langbourne Williams, Jr. replaced Eric Swenson as Freeport's chief executive. Williams' first task was the development of the Grande Ecaille property, which would require spending millions of dollars just to bring equipment to the site, which was more water than land. The risk was such that to secure financing, the company had to put up Hoskins Mound as collateral. By January 2, 1933, construction had begun in earnest, and eleven months later, on December 7, the first wells were steamed. During the very first month, daily production averaged 708 tons, meeting the design yield of 700 tons per day. Eventually, the production rate reached 1,200 tons per day, which delighted Freeport's managers and investors.

When reviewing the records of past construction programs such as Bryanmound, Hoskins Mound and Grande Ecaille, one is impressed by the speed with which these projects were carried out. This was before computers, cellular phones, critical path scheduling, and the numerous productivity enhancements of today. Yet major construction programs were completed, in very short periods of time, with what today would be considered primitive equipment.

Although government permits, environmental impact statements and other modern paperwork contribute to delays today, we are perhaps also slowed by our preconceived ideas about how a job should be done. For example, when we drive piling today, we assume that the tops of the piles must be cut off with a chainsaw. A chainsaw is a dangerous tool, and therefore before it is used the site must be prepared to allow for stable footing. The photograph on the next page shows how it was done before modern tools. A couple of men would wade in with a cross-cut saw and start cutting. Although a chainsaw would seem to be a much more efficient instrument, in fact two men with a handsaw today could probably go in and finish the job even before the site could be prepared for the chainsaw.

Transportation and logistics provided the major obstacles to developing Grande Ecaille. To solve the transportation problem, Freeport bought an industrial town site on the Mississippi River that was served by railroad and state highways. This town, 45 miles southeast of New Orleans on the Mississippi River, was to become known as Port Sulphur. To provide rapid and economical transportation from the Mississippi to the mine, a 100-foot-wide, 10-mile-long canal was built from Port Sulphur to the mine. Port Sulphur soon evolved into a thriving residential and industrial community, complete with hospitals, schools, recreational facilities, and golf courses for the employees. The attractively laid out town was the model industrial community of its time. Port Sulphur today is a community of 3,500 merchants, fishermen, and retired and active Freeport employees, and Freeport Sulphur still keeps its terminal facilities there.

Freeport Sulphur workers, in 1933, cut off pilings the old-fashioned way. The men are preparing a foundation for the facilities at Grande Ecaille.

While Grande Ecaille was being developed, the company became interested in minerals other than sulfur, and in the potential of overseas operations. Freeport acquired a controlling interest in the Cuban America Manganese Corporation which, while never economically important, provided an entrée into other operations in Cuba. By 1940, the Nicaro Nickel Company was chartered as a subsidiary of Freeport Sulphur for the production of nickel from Cuban ores.

During World War II the demand for Freeport's commodities increased. Annual sulfur production grew to one million tons and the company, with a contract to supply nickel to the United States government, began construction in Cuba of a full-scale plant to produce this scarce resource. The nickel plant was based on a newly developed process of extracting nickel from lateritic ores. By the time this operation was shut down in 1947, it had contributed 63 million pounds of nickel to the war effort.

During the 1940s methods of sulfur production began to change. Where in the past sulfur had been shipped as lumps of solid sulfur, Freeport began to experiment with, and finally switch to, the delivery

Port Sulphur was a model company town in its day. Freeport Sulphur still maintains sulfur shipping facilities here. Photograph from 1952.

of liquid sulfur. Because it eliminates the corrosive dust associated with solid sulfur, this technique was considered one of the major environmental breakthroughs of the 1940s.

In the mid-1950s, Freeport's mining and production facilities were in Texas, Louisiana, and Cuba, and its corporate headquarters were in New York City. In this same decade, the company expanded into potash, an important raw material for fertilizer, by buying 50 percent of the National Potash Company, with a mine near Carlsbad, New Mexico. In Louisiana, Freeport brought the Chocagoula, Caminada, and Grand Isle sulfur mines into production. The company's interest widened into oil, and in 1956 chartered the Freeport Oil Company. Oil was a natural development, as drilling and geological technologies were essentially the same in the sulfur and oil industries, and the Gulf Coast zone in the 1950s was of importance to both. Salt domes, where sulfur is occasionally found in commercial quantities, were of geological importance to petroleum geologists.

During this decade Freeport chartered the Moa Bay Mining company to produce nickel and cobalt from ores at Moa Bay, Cuba. To refine the nickel sulfides mined in Cuba, Freeport built a smelter on the Mississippi River near New Orleans. This was a state-of-the-art facility, and the first hydrometallurgical smelter. It converted nickel sulfide to nickel metal by chemical means, thus avoiding the plumes of smoke produced by pyrometallurgical smelting. Construction and development of the Cuban nickel operations were a major highlight for the company, but quickly brought perhaps its biggest disappointment. In 1959, just after the company shipped its first boatload of nickel, Fidel Castro's government expropriated the Moa Bay mine. This mine still operates today under Cuban government management, producing about 4 percent of the world's nickel. The smelter went to the banks, as collateral for loans unpaid because of the expropriation, and it is today operated by AMAX to refine waste materials.

In 1960 Freeport's Grand Isle project began production, becoming the world's first offshore platform sulfur mine. The use of full-strength sea water in this mine was a major technical achievement, and a first for the industry. By 1963, Freeport's production reached 2.5 million tons a year, making it the world's largest sulfur producer.

The loss of Freeport's nickel operations triggered further worldwide diversification. By the 1960s, exploration began in Australia, with the company's geologists looking primarily for nickel. Forbes Wilson was actually looking for nickel prospects in Indonesia when he encountered a report of copper in the interior of what was then still Netherlands New Guinea. His 1960 expedition to the Ertsberg copper deposit was a direct result of the company's worldwide search for nickel deposits to replace its lost Cuban operations.

The company continued to grow and diversify in the 1960s. In 1964 it purchased Southern Clays, Inc., which was expanded and renamed Freeport Kaolin Company. In the late 1960s, with the changing leadership in Indonesia, work on the Ertsberg mine began. In 1969 Langbourne Williams, chief executive of the company since the 1930s, retired and was succeeded by Thomas Vaughan. By 1971, Freeport's activities had become so diverse that the company's name was changed to Freeport Minerals Company. In 1972 the Greenvale nickel cobalt project in Australia began production, and the first shipment of Ertsberg copper concentrate left Indonesia. In 1975, Freeport's top leadership changed again, as Paul Douglas replaced Thomas Vaughan as president and chief executive of the company.

The 1970s brought depressed metals prices and soaring fuel

costs, which severely stressed the Freeport Indonesia and Greenvale projects. In both cases, reasonable fuel prices were critical to the profitability of the operations. With shortages and high prices in the oil industry, Freeport Oil stepped up its operations, and Freeport Minerals began buying an interest in McMoRan Oil and Gas. In 1981, the two companies combined, forming Freeport-McMoRan Incorporated. Gold prices were also high during this period, and in 1981 the company formed Freeport Gold, and began the Jerritt Canyon gold operations in Nevada. By 1982, Freeport's Nevada operation was the single largest producer of gold in North America.

The 1980s became an era of restructuring. The diverse nature of Freeport's operations made it necessary to concentrate the company's energy in areas where growth was likely. To accomplish this, the top management structure was reorganized. In 1984, Jim Bob Moffett became chief executive and chairman of the board. An "office of the chairman" was created, with Milt Ward as president and chief operating officer and Nils Kindwall as co-chairman and chief financial officer. Though the individuals have changed, the general concept of an office of the chairman with an operating officer and a financial officer remains to the present day. The restructuring also included the sale of Freeport Kaolin, and the consolidation of Freeport's management staff into new headquarters on Poydras Street in New Orleans.

By 1986, the new Freeport management became concerned with the diminishing ore reserves of its many subsidiaries, and a mandate was issued to all divisions to find new reserves. Each division was held responsible for enlarging its asset base, and producing a foundation for additional growth.

Freeport Indonesia was fortunate in that the contract area around the mine was a productive location for minerals exploration. Additional nearby underground ore reserves were discovered, as well as Grasberg. When the drill results first came back from Grasberg in 1988, the possibilities looked very good, but it was by no means a proved deposit. Still, in 1988 the company formed a New York stock exchange company, Freeport-McMoRan Copper & Gold (FCX on the ticker), which would own all of Freeport's assets in Indonesia.

The 1980s brought further diversification and expansion in sulfur and agricultural chemicals. In the 1960s Freeport had begun a downstream diversification, establishing Freeport Chemical Company and building a phosphate plant at Uncle Sam, Louisiana. The plant processed phosphate rock from Florida with Louisiana-produced sulfuric acid to produce phosphoric acid. This, in turn, was used by the Agrico Chemical Company to manufacture high-analysis fertilizer for sale to the corn belt of the Midwest and to farmers worldwide. In the late 1980s, this operation found itself with insufficient volume to achieve good economies of scale. This problem, together with its generally uncertain financial performance, was solved by buying Agrico's phosphate and nitrogen assets from the Williams Company, thus expanding Freeport Chemical's volume, product mix and sales distribution systems.

The same year as the Grasberg discovery, 1988, was also a memorable one for the sulfur and agricultural chemical group as well. A remarkable sulfur deposit was discovered at Main Pass Block 299 in the Gulf of Mexico, some 20 miles from the mouth of the Mississippi River. Main Pass and Grasberg, coming at the same time, required a new look at the company's structure, and at what should be done to best develop these exciting new discoveries.

The company's main concern in the late 1980s was to find the money for the massive $1.4 billion Main Pass sulfur project. This required a close look at the company's assets. All, in their own right, were successful, but in the end Freeport sold the Jerritt Canyon Gold Company, various sectors of Agrico, the Australian production and minerals exploration operations, and many oil and gas properties.

The nineties began a new period for the Freeport family of companies. Freeport Indonesia continued to expand almost exponentially, until by mid-1995 the amount of ore processed reached 125,000 metric tons per day, at the same time moving more than 500,000 tons per day of waste. At this capacity, the Indonesian operations produced approximately 5 percent of the world's newly mined copper, and the mine itself was among the very largest in the world. In effect, the Indonesian child had now outgrown the parent.

The investment and financial community saw potential conflicts for stockholders in continuing the organization of Freeport-McMoRan Copper & Gold—with its rapidly expanding Indonesian property, P.T. Freeport Indonesia—as a subsidiary of Freeport-McMoRan. Banks that could lend to an Indonesian copper operation faced restrictions lending to a company in the fertilizer business, and investors interested in fertilizer stocks did not want an investment

MAP 1.1

that also involved copper and gold. Analysts who understood the metals business did not understand fertilizer, and vice versa. Therefore, in 1995, Copper & Gold stocks were distributed to the mother company, and each of the two companies proceeded on their separate courses. At the time of this writing, there is still some shared management and facilities, but this will end over time.

INDONESIA IS THE least well-known of the world's largest countries. It ranks fourth in population, with approximately 200 million inhabitants. There are more Muslims in Indonesia than anywhere else. The country's area, about the size of Mexico, covers just over 2 million square kilometers of dry land, but the island nation stretches more than 5,000 kilometers east to west along the equator, about the distance from New York to San Francisco. It is a developing country with a slowly rising standard of living. Most of the population is rural, although many now seek their fortunes in the cities. The nation's capital and largest city, Jakarta, on the island of Java, is home to 9.3 million people. Most of Indonesia's population is concentrated

33

on Java (about 60 percent) and the other islands of western Indonesia, and the rest of the country is very sparsely populated.

President Suharto has led the Republic of Indonesia since 1967, and the calm political climate established under his leadership has allowed steady economic development. Indonesia is an Organization of the Petroleum Exporting Countries (OPEC) member and major oil producer, but oil exports have declined recently. Natural gas, plywood, copper, coal and clothing are now accounting for an increasingly large share of export revenues. Tourism has become the third leading source of foreign exchange, and might well become the leading one within a decade. Thanks to improving infrastructure, foreign visitors are exploring parts of this vast country beyond Java and Bali.

Indonesia includes several of the world's largest islands, and thousands of smaller ones. The last official survey counted 18,585. Working across the archipelago from the east, the nation begins with Irian Jaya, a province covering the western half of the Island of New Guinea, the world's second-largest island, which lies just north of Australia. Next comes Maluku, also called the Moluccas or Spice Islands, a dispersed group of mostly small islands including those that until well into the 1800s were the world's only source of cloves, nutmegs and mace. Just west of Maluku and due south of the Philippines is Sulawesi, once called the Celebes, its distinctive gangly shape the product of the tectonic collision of two island masses. South of Sulawesi are the Lesser Sundas, an island chain that stretches from Timor in the east to Bali in the west. West of Bali are Java, and just past tiny Krakatoa volcano, Sumatra, the world's fifth-largest island. North of Java are the four provinces of Kalimantan, which cover about two-thirds of Borneo, the world's third-largest island.

The first of our species to inhabit these islands are thought to have arrived 30,000–40,000 years ago—according to some estimates, as early as 60,000 years ago—from Africa, taking advantage of the lower sea levels to make their way as far as Australia. Little is known of these groups, and there seems to have been more than one migration. Among these were the ancestors of the people currently living in New Guinea.

The ancestors of most of today's Indonesians started out in south China, some 5,000 to 7,000 years ago, soon reaching what is now Taiwan. There, at about the same time the wheel was invented in the Middle East, these people—whom anthropologists call the Austronesians—came up with two inventions of equal importance: the sail and the outrigger. Boats equipped with sails and outriggers could cross relatively wide stretches of water, and in small-scale migrations lasting hundreds of years, they fanned out across the South China Sea to what is now the Philippines and Indonesia. Later, some of the Austronesians made it all the way to Easter Island, and a few remain in Taiwan. Linguistic work with these aboriginal Taiwanese, now far outnumbered by late-arriving Chinese, has allowed anthropologists to reconstruct the migration patterns.

The Austronesians brought with them pottery-making technologies, and methods for growing irrigated rice and raising domesticated animals. They worshiped the spirits of their ancestors and those of their new homes, and spread throughout the Indonesian archipelago from about 3000 B.C. to 500 B.C. Around the time of the birth of Christ, elements of the civilization of India began to reach western Indonesia. Brahmin priests endowed local leaders with the divine right to rule, and kingdoms grew to become empires.

The first of Indonesia's great pre-colonial empires was the Sriwijaya, a Buddhist maritime empire based near present-day Palembang in south Sumatra. Its strategic position allowed the empire to control the Straits of Malacca, the route of the lucrative trade with both China and India, from the 7th to the 11th century. Sriwijaya was followed by empires on Java, based on large surplus rice cultivation, which resulted in the remarkable temples of Borobudur and Prambanan, in central Java. The golden age of classical Indonesia was reached with the rise in East Java of the Hindu Majapahit Empire, the most powerful the archipelago had known. At its height, in the mid-14th century, Majapahit claimed suzerainty over more land than makes up Indonesia today.

The fall of Majapahit is often claimed to have come from a defeat by the Muslim Demak kingdom in northern Java in the 15th century. Although scholars consider this account apocryphal, the Arab presence in the archipelago was by this point considerable. Arabs much earlier had established a monopoly on the trade in spices to the Middle East and Europe, and the traders spread Islam in Sumatra, Java and the Spice Islands.

Although in the late 13th century Marco Polo stopped for several months in northern Sumatra, other early European tales of travel in what is now Indonesia are fabrications. The first westerners to make

35

it to the Indies, as they were then called, were the Portuguese. In the 15th century, they gradually explored further and further down the coast of Africa, rounding the Cape of Good Hope in 1489. They soon reached India, and landed at the entrepôt of Malacca, on the Malay peninsula, in 1511.

The Portuguese sought the famed spices of the East, which were until then brought to Europe by Arab traders—who kept the source of their wares secret—and sold for a very high price. The Portuguese discovered that pepper, traded in Malacca, came from just across the strait in Sumatra, and cloves and nutmegs were brought in from the Spice Islands. The following year the Portuguese found the way to the clove-producing islands of Ternate and Tidore, and for the next century enjoyed a near monopoly on the lucrative European trade in pepper, cloves and nutmegs.

The Dutch began arriving in the Indies at the very end of the 16th century. With more maneuverable ships and far better organization, Protestant Holland soon took over the spice trade from the Catholic Iberians. The Dutch organized a trade consortium, the Vereenigde Oostindische Compagnie or VOC, better known in English as the Dutch East Indies Company. First through the VOC, and later through a colonial government, Holland controlled the East Indies for the next three and a half centuries from Batavia, now Jakarta.

European rule ended in the East Indies with World War II. In 1942, Japanese forces swept through the Dutch Indies, making short work of the outnumbered and outgunned Dutch colonial army. When the war ended three years later in an Allied victory, Holland—with British aid—attempted to reclaim her colony. But two nationalist leaders, Sukarno and Mohammad Hatta—both of whom had been previously jailed by the Dutch—had declared Indonesian independence on August 17, 1945, even before the Japanese formally surrendered. But it was only after a nasty, four-year war, and repeated threats by the United States to cut off Holland's Marshall Fund money, that the Netherlands government withdrew and acknowledged their former colony's independence.

When independence was won in late 1949, Sukarno became Indonesia's first president, and Hatta the nation's vice president. Sukarno's Indonesia became a leader of the non-aligned movement, but his increasingly ineffective economic policies and disastrous policy of military confrontation with Malaysia in North Borneo began

Art adorning coastal cliffs near Kaimana and Fakfak—this example is from Bitsyari Bay—suggests a connection between Papuan and Aboriginal Australian culture.

to erode his support. Sukarno's politics moved steadily to the left, and he grew hostile to the West and increasing friendly to China and the Soviet Union. A master politician, he long played the country's four main political factions against one another: the Communists, the military, the nationalists and the Muslims.

In 1965, the Indonesian Communist Party (PKI) attempted a coup, which was quickly suppressed by the military. This set off a national reaction, in which hundreds of thousands of people were killed as suspected Communists. Sukarno quickly fell from power after the coup, and General Suharto took over. In 1968 Suharto was formally elected president, and he has been regularly reelected ever since. Suharto inaugurated the "Orden Baru," or New Order, emphasizing economic advancement and friendly relations with the West.

Two of Indonesia's 27 provinces were added after independence. In 1963, Dutch New Guinea fell under Indonesian control, and later became the province of Irian Jaya. In 1975, following a sudden, worldwide decolonization by Portugal, the eastern half of the island of Timor was invaded by Indonesian troops, and Indonesia annexed East Timor in 1976. This annexation is not universally recognized by the international community, and is a recurring source of criticism.

Almost 90 percent of Indonesia's population is Muslim. There are scattered Christians and Buddhists, and a few million Hindus, these almost exclusively on Bali. Roman Catholicism remains the dominant religion on Flores Island and in East Timor, where the Portuguese influence was strongest in the past. More recently, Catholic missionaries have successfully converted people in parts of Maluku, southern Irian Jaya, and the interior of Kalimantan. Protestant Christianity is strongest in Central Maluku, north Sulawesi, the highlands and north of Irian Jaya, and West Timor. Traditional belief systems are still very strong in Irian Jaya, and among the Toraja of Sulawesi, the Bataks of North Sumatra and the Dayaks of Central Kalimantan.

F REEPORT'S MINING AND exploration activities are concentrated in Irian Jaya, the most remote part of Indonesia and, arguably, the most remote and least known place in the world. New Guinea is the largest island in the world after Greenland, covering 792,540 square kilometers. The island measures 2,400 kilometers east to west, and at its widest point, in the middle, 740 kilometers north to south.

New Guinea's current political division is the result of past colonial claims. The eastern half, formerly British and German territories, has since 1975 been part of the independent nation of Papua New Guinea. The western half, formerly Dutch-claimed, is the Indonesian province of Irian Jaya. The border between the two countries runs neatly along a line following 141 degrees East longitude, except for a slight blip along the Fly River, just north of Lake Murray. Irian Jaya covers 421,981 square kilometers, roughly the area of California, which represents 22 percent of Indonesia's land area. But only about one percent of the nation's population lives in this province, which is still mostly wilderness.

A towering limestone cordillera rears up for much of the length of the island, forming a backbone of 3,000-meter peaks. The mountain chain is fringed by alluvial plains to the north and south. The highest mountains are on the Indonesian side, crowned by Puncak Jaya, just five kilometers from the Grasberg mine. Reaching 4,884 meters, it is the highest peak in Southeast Asia. The entire mountain range shows signs of mineralization, especially copper and gold.

New Guinea's first gold rush took place near Port Moresby, on the eastern side, in 1878. Because of extensive British and Australian explorations, the mineral resources east of the border have been mined for several decades. Freeport was the first—and so far the only—company to mine in Irian Jaya.

Approximately 1.9 million people live in Irian Jaya, a bit over 1 million of whom are Papuans, or aboriginal Irianese. They are among the most traditional people remaining in the world, and in all but a few of the coastal areas, Stone Age conditions prevailed until just a few decades ago. Many highlands Irianese still wear penis gourds and grass skirts as their only articles of clothing. Tribal warfare is no longer common, but it is by no means extinct, and traditional agriculture is still the norm.

This one province contains roughly 40 percent of Indonesia's languages, 251 out of slightly fewer than 600. The languages of Irian Jaya, combined with the 770 languages spoken in Papua New Guinea, represent fully one-fifth of the world's languages. Some 140 of Irian's languages are spoken by fewer than 1,000 people. One-third of the province's indigenous people speak one of the Dani (230,000) or Ekagi (100,000) dialects, and only these two groups and perhaps the Asmat (60,000) of the south coast are very well known.

Humankind arrived in New Guinea long before it reached the

Dugout canoes, such as this Kamoro example, are handy in narrow tidal channels, but it was the invention of the outrigger that brought Austronesians to the islands.

Americas or most of Europe. Circumstantial evidence suggests that 30–40,000 years ago the ancestors of today's Irianese arrived in small-scale migrations that coincided with periods of glaciation, and consequently lower sea levels. The island that greeted these pioneers was drastically different from the one we find today. Great sheets of glacial ice covered extensive areas, and the snow line was just 1,100 meters above sea level. Average temperatures were 7 degrees Centigrade cooler than today.

During the various ice ages, the shallow Arafura sea became a land bridge, and the populations of aboriginal Irianese and aboriginal Australians mixed. Evidence of this can be found in the styles and motifs of the rock art along Irian's southwest coast and Arnhem Land in northern Australia (see photograph page 37). As the climate warmed, the aboriginal Australians adapted themselves to a different, and much drier ecology. Genetic evidence suggests the two populations separated some 10,000 years ago.

The Austronesians were the last major group to migrate to Irian, arriving by sail-powered and outriggered canoes between 4,000 and

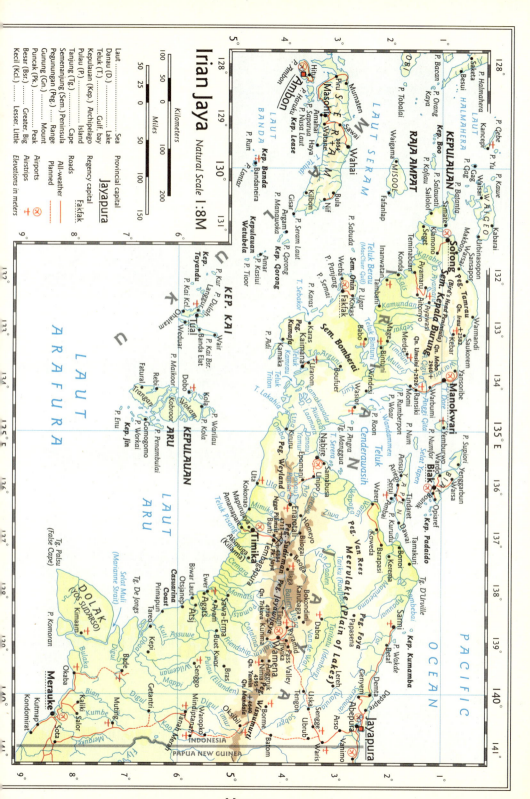

Map 1.2

5,000 years ago. They brought with them a high level of social organization, characterized by bilateral descent affiliation—tracing one's ancestry through both the father's and the mother's clans. The Papuans of Irian, much like Europeans and Americans, trace descent only along their father's line. The Austronesians also had superior technology: better weapons, agricultural implements and tools, and domesticated animals. Elsewhere in Indonesia the Austronesians displaced, and finally replaced, the original inhabitants. In New Guinea, they never penetrated beyond some offshore islands and scattered areas along the north shore.

By the time the Austronesians arrived, the aboriginal Irianese had developed into well-established agricultural groups, with enough social cohesiveness to maintain their cultural and linguistic identity. By at least 4,000 B.C., the Irianese were cultivating taro, a tuber crop, and keeping domesticated pigs. The later introduction of the hardy and nutritious sweet potato led to the opening up of the fertile highlands. This plant, *Ipomea*, is of New World origin and how it got to the interior of Irian Jaya is still a mystery. This crop allowed high population densities to develop in the fertile, 1,500-meter-high valley of the Baliem, out of reach of the worst of the malarial mosquitoes.

Trade between New Guinea and central and western Indonesian probably started before the Christian era. Traders from the western part of the archipelago brought cloth and metal implements to exchange for bird of paradise feathers, massoi bark—used in Java to prepare medicines—and slaves. The frizzy-haired people represented on some of the friezes on the 8th-century Buddhist temple of Borobudur were very likely Papuan. The Negarakertagama, a 14th-century poem of the Majapahit kingdom in east Java, tells us that two Papuan territories were part of the great empire. And well before the arrival of the first Europeans, the rulers of the tiny island of Tidore in Maluku claimed areas of western New Guinea. This clove-producing sultanate sent both trading and raiding expeditions to New Guinea.

Thanks to written records, the history of European contact with the people of New Guinea is much more complete. Because of their clove-trading outposts in Maluku in the early years of the 16th century, the Portuguese were the first Europeans to hear about New Guinea. An early governor of the Portuguese Moluccas, Jorge de Meneses, blew off course in 1526 on the way from the Malay peninsula to his new domain and landed on the north coast of New Guinea's large western peninsula, called the Bird's Head. Forced to wait for the winds to change before heading back west, he was the first European to describe New Guinea, which he called "Ilhas dos Papuas." The origin of the word "Papua" is unclear; although most sources say it comes from old Malay, it is more likely to have come from a word in one of the Moluccan languages. Sailing along the island's north shore in 1545, on his way back to Mexico from the Moluccas, the Spanish captain Yñigo Ortiz de Retes named the island Nueva Guinea—probably because he mistook the inhabitants for Africans. This designation first appeared on Mercator's world map of 1569.

The Dutch appeared in the Indies in the early 17th century, and in 1606 the captain and navigator Willem Jansz, looking for gold, touched land on New Guinea at various points. In 1623, Jan Carstensz, sailing along the south coast, sighted snow-capped peaks in the vicinity of what is now Freeport's mine area. His report was ridiculed in Europe, as glaciers in tropical latitudes were then considered an impossibility.

During the 18th and 19th centuries, some of the greatest names in the history of navigation and exploration sailed by and sometimes briefly stopped in what is now Irian: William Dampier (1700), Louis Antoine de Bougainville (1768), James Cook (1770), and Dumont D'Urville (1827). None stayed long. The coasts were densely forested and malarial, and the Papuans were definitely not friendly.

The first attempts at European settlement were defeated by disease, hostile Papuans, and the inability to find economic quantities of spices or other trade goods that had a European market. In 1775, Thomas Forrest of the British East India Company, looking for a source of spices outside the Dutch sphere of influence, landed at Dorei Bay, near present-day Manokwari, and learned of the traditional trade in massoi bark, bird of paradise plumes and slaves. He thought this commerce unworthy of the company's attention.

Captain John Hayes was not so discouraged, and in 1793 set up a small colony at the bay Forrest visited, rather hubristically claiming the whole of New Guinea for England, and renaming it New Albion. After leaving his settlement in what he thought was good order, Hayes became caught up in political intrigues between England, Holland and France—at this time, Holland and England had declared war on revolutionary France. In his absence, many of the colonists died of disease or fell in skirmishes with the local Papuans. Some

Jan Carstensz, when he first reported this sight in 1623, was ridiculed in Europe, as it was thought impossible that glaciers could form in the tropics.

were taken slaves. A year-and-a-half later, the remaining colonists, all of them ill, were evacuated.

After the Napoleonic Wars, the Dutch sent a brig under Lt. Dirk Hendrik Kolff to rekindle trade contacts with the people of the southeastern Moluccas, and to check on the condition of old Dutch outposts there. During this journey, undertaken in 1825–1826, Kolff explored and briefly stopped on the south coast of Irian Jaya. His report led to the Dutch claims to the south coast of New Guinea up to the 141st parallel. Holland followed up this claim by establishing Merkusoord, a colony on Triton Bay, but the ravages of malaria forced the abandonment of the colony in 1835. A few scattered stones from the settlement's Fort du Bus—named after the Belgian nobleman Viscount du Bus de Ghisignies, then governor-general of the Indies—can still be seen.

Holland abandoned further attempts at settlement, but in 1848 defined the limits of its western border as longitude 141 degrees east in the south, and longitude 141 degrees 47 minutes east in the north. The north coast of this area was formally placed under the Sultan of

Although a barrier to exploration, Irian Jaya's lowland swamp forest is rich in wildlife and plant life, including the staple sago palm and the pandanus, seen here.

Tidore, by that time a vassal of the Dutch. In 1884, from its base at Port Moresby, England claimed a large chunk of the southeast part of the island, and the following year the German Imperial flag was raised over the northeast sector. It wasn't until 1898 that the Dutch established posts at Manokwari on the north coast, and Fakfak, an old trading center, on the south. In 1902, responding to British complaints that Dutch subjects were conducting head-hunting raids across the border, the Dutch colonial administration established Merauke, on the south coast near the border with British New Guinea. By the early 20th century, the de facto boundary at longitude 141 degrees east had been settled by the European parties.

The Dutch colonial administration was at first loathe to expend resources on the large, unmanageable and economically unrewarding territory of western New Guinea, but in the 19th century the unexplored island attracted the attention of several naturalists. Alfred Russell Wallace, whose biological explorations of the East Indies led him independently to a theory of the history of speciation much like Charles Darwin's, spent five months in 1858 in the area of Dorei Bay.

The Russian naturalist Nikolai Miklouho-Maclay spent three months on the southwest coast of Irian in 1874, making a brief foray to Lake Kamakawalar, just inland of Triton Bay. Two years later, the Italians Luigi D'Albertis and Odouardo Beccari trekked into the Arfak mountains, inland from Manokwari. In addition to collecting specimens, all these early scientist-explorers wrote briefly about the Papuans, providing the first written records of the Irianese.

During the opening years of this century, the Dutch finally began exploring Netherlands New Guinea in earnest. Between 1900 and 1930, the colonial government sponsored a total of 140 expeditions. The most important of these were the huge military forays inland between 1907 and 1915. By this point, the major river systems and basic geographical features of the island were mapped. The last major discovery occurred in 1938, when an expedition led by the American explorer Richard Archbold landed a seaplane on Lake Habbema and discovered the Baliem Valley. The Grand Valley of the Baliem, a fertile alluvial plain 50 kilometers by 15 kilometers, then supported 50,000 Dani, who had no contact with the outside world until Archbold. Today this valley is Irian's prime tourist destination.

Missionaries also were among the first explorers. In 1852 two German preachers, recruited by the Dutch Protestant Church, established a mission at Dorei Bay. Life was no easier for them than for the Europeans who previously had tried to settle these shores. The mission hung on for a quarter of a century but made few converts. More missionaries died of malaria than Papuans were converted. As Dutch administration began to take hold along the coasts, more and more missionaries arrived. To keep religious rivalries in check, the Boundary of 1912 was created, restricting Protestants to the north, and Catholics to the south, the same as in the Netherlands. This boundary accounts for the distribution of the two faiths in Irian Jaya today, although the Boundary was abandoned in 1955.

After World War II, American Protestant missionaries began to arrive. These men and women, financed by churches in the American Midwest, were of a more fervent, evangelical bent, and they braved the rugged highlands, working their way east from their first base in the Paniai Lakes region. They used medicines, steel ax heads, valuable beads and tobacco as inducements to convert their charges. The results of their efforts can be seen today—many of the highland Irianese are New Testament Christians.

As elsewhere, the work of the missionaries in Irian Jaya has come in for its share of criticism. The evangelical Protestants in particular expected quick results, but the traditional beliefs of generations cannot be broken with overnight. Massive burnings of "fetish" carvings and outright bribery with tobacco and other goods were among the excesses of the early days.

The introduction of the new faith, combined with the obvious political and economic power of the missionaries, also helped foster "cargo cult" thinking. Cargo cults are pseudo-religious movements, often with a millenarian cast, that have sprung up throughout the Pacific Islands wherever Europeans, and their 20th-century goods, first met traditional Melanesian cultures. These cultures believed that since it was the spirit world that provided good health and material possessions, the westerners' medicines, radios and other "magical" technologies must be the result of some secret ritual knowledge. Particularly in the context of the introduction of a new religion, this thinking led to jealousies, resentment and local rebellions.

Despite scattered colonial efforts and the work of the missionaries, New Guinea was, to most people, unknown until World War II. The Japanese swept through the Dutch Indies in 1942 almost unopposed by the Dutch colonial forces. The forces of the Imperial Army were stopped only just outside of Port Moresby—and held there—by a group of tough Australian troops.

By the spring of 1944, the United States had entered the war and prepared the troops and matériel to attempt a counter-offensive. Under the leadership of Gen. Douglas MacArthur, parts of the northeast coast of New Guinea were retaken. Then, thanks to intercepted Japanese communications and broken codes, the defensive weakness of Hollandia became known. Hollandia was by then the principal town of Netherlands New Guinea and held an airfield at nearby Sentani. The Japanese had gathered 11,000 men there, but only one-fifth were combat troops. With a powerful force of 1,200 planes, 217 ships and 50,000 troops, MacArthur landed on Hollandia, securing the town at the cost of only 159 Allied soldiers.

Hollandia—now Irian Jaya's capital city of Jayapura, with a population of 310,000—quickly grew into a great military base, with 250,000 soldiers and support personnel. The north coast of Dutch New Guinea soon fell to the Allies, but it required a bloody effort to overcome the 10,000 Japanese troops on Biak Island, entrenched in a

A World War II gun emplacement at Keakwa, 20 miles west of Freeport's Amamapare port facilities. The eroding coastline now exposes the gun on the beach.

system of interconnected caves. The battle for Biak was among the toughest fought in the war, and American troops resorted to such measures as using dynamite and diesel fuel to drive the Japanese from their caves. Only 220 Japanese soldiers survived, and families of the veterans still come to Biak to honor their fallen relatives. The Allies then extended Biak's airfield to accommodate the long-range Superfortress bombers that turned the tide of the Pacific war.

After World War II, and a four-year independence struggle with Holland, the new nation of Indonesia was born. Sukarno, the first president of Indonesia, had always considered Netherlands New Guinea to be part of the former colony that should be incorporated into the new nation. But right-wing forces in Holland, stung by the loss of their East Indies empire, clung to Netherlands New Guinea. To bolster her case, the Dutch government, for the first time, began to pour resources into New Guinea. Sukarno never abandoned his fight, however, and finally pulled Indonesia out of the United Nations, nationalized Dutch businesses, and embarked on a military

Freeport's mine access road is just one example of the infrastructural improvements the company has brought to its contract area in southern Irian Jaya.

campaign to drive Holland from west New Guinea. Under the leadership of a young general named Suharto, paratroopers were dropped inland and naval engagements took place off the coast.

The Dutch were able to neutralize Indonesia's military actions, but world opinion was changing. The United States, fearing that a protracted struggle would push Indonesian even further into the Soviet camp, decided to support Sukarno's claim to Netherlands New Guinea. To smooth the transition, the territory was passed to United Nations control in 1962, and the next year given over to Indonesia. This transfer carried the stipulation that a referendum—an "Act of Free Choice"—be held before 1969 in keeping with the United Nations' stated principles of self-determination. The Indonesian government chose 1,000 chiefs, who voted unanimously to join Indonesia. In 1973, at the dedication of Freeport's mining town of Tembagapura, President Suharto renamed the province Irian Jaya.

Indonesian administration has brought considerable progress to Irian Jaya. Despite the severe handicap of rough terrain and almost non-existent roads, schools have been built in almost every village area, no matter how remote. The larger villages and cities now have clinics and hospitals. Most Irianese have entered, if only in a modest way, the monetary economy. Air communications within the province, and with the rest of Indonesia, have been established. Many coastal towns are connected by efficient passenger ships and freighters. More and more Irianese are being integrated into the local administration, and jobs and business opportunities are becoming available to at least some of them. Bahasa Indonesia, the national language, is now a second language for younger Irianese, and many of the older generation, except in the most remote areas.

Progress has not been entirely smooth, however. Before they left Netherlands New Guinea, the Dutch encouraged anti-Indonesian sentiment and fueled a desire for independence. This desire, and excesses in the early Indonesian administration of the province, led to the formation of the Organisasi Papua Merdeka (OPM) or Free Papua Organization, a small political and military group opposing Indonesian rule. With the cooperation of neighboring Papua New Guinea, which no longer allows OPM activity on its soil, the Indonesian military has mostly suppressed the movement. But occasional actions still take place, and localized resentment continues, brought on by the visible disparities between the material conditions of the tribal Irianese, and those of recent immigrants from western Indonesia.

The Indonesian government realizes that it will take education, understanding and an improving provincial economy before these resentments disappear altogether. The controversial transmigration program, a scheme begun under the Dutch and financed by the World Bank to relocate Indonesians from crowded Java and Bali to the more remote parts of the archipelago, has been scaled back, and is handled more thoughtfully than it has been in the past. Early mistakes led to settlers being placed on traditional hunting grounds, or in areas unsuitable for rice cultivation. Today, transmigration sites are carefully selected to account for traditional ownership, soil type, and the local availability of infrastructure and marketing potential.

Indeed, one of the province's most successful transmigration areas has been around the town of Timika, where the presence of the Freeport operations have created an efficient basic infrastructure, as well as a lucrative market for locally produced vegetables and other agricultural surpluses.

CHAPTER TWO

The Discovery of Ertsberg

A shocking visit—Terra incognita—British ornithologists—The race to the snow—Oil exploration—The Colijn expedition—Ertsberg—A forgotten report—Forbes Wilson—Gleaming chalcopyrite—13 acres of ore

"IN 1959 THIS American visits me and says, 'Say, you were in New Guinea and found this ore body. How big is it?' After twenty-three years the shock of this question was like somebody holding a pistol to my chest. 'Well,' I said, 'it was a wall'—and I didn't want to exaggerate—'about seventy-five meters high, and something similar in length.' 'Oh, oh!' was his reply. He was the first person to go see for himself whether or not I was talking nonsense. I met him again when he came back, and he said, 'It was much bigger than you said!'"

This is Jean Jacques Dozy's recollection of his first meeting with Freeport Sulphur exploration manager—and later Freeport Minerals president—Forbes K. Wilson, the first person to see potential in a geological anomaly Dozy noted during an expedition to the Carstensz mountains in 1936. In his notebook, Dozy sketched a strange, squat outcrop of black rock standing at the foot of an alpine meadow 3,500 meters high in the interior of New Guinea. Underneath it he wrote "Ertsberg," which is Dutch for "Ore Mountain."

Dozy, at the time a young petroleum geologist, was a member of the 1936 Colijn Expedition, the first group to successfully reach the cluster of peaks around Puncak Jaya, which were then called the Carstensz Toppen.

At the turn of the 20th century the outline of Netherlands New Guinea was pretty well mapped, but the interior was almost entirely terra incognita. A prize that particularly tantalized and frustrated explorers were the snow-clad mountains near the current Freeport operations, which, when the weather cooperates, are visible from the southern coast.

Captain Jan Carstensz, sailing along the south coast of New Guinea, noted in his log for February 16, 1623: "[W]e saw a very high mountain range in many places white with snow, which we thought a very singular sight, being so near the line equinoctial." By the beginning of the 20th century, Carstensz's mysterious mountains, and the unknown interior of New Guinea in general, were the last major blank on the world's map.

Although they had been in the Indies since the early 17th century, the Dutch paid almost no attention to New Guinea until the very end of the 19th century. The island, though vast, held little of economic importance to the Dutch, and it was too far away—and in some areas the Papuans were too violent—to govern cost effectively. But in the 1870s, Germany began looking for Far East colonies, and in 1883 Bismarck claimed the northeastern section of New Guinea for Germany, dubbing it Kaiser Wilhelmsland. This spurred the Dutch to strengthen their claims to the western half of the island, which had existed up until this time only through their vassal, the Sultan of Tidore. The Dutch-British boundary was settled in 1895 (the Dutch-German boundary was settled in 1910) and in 1898 the Dutch established posts at Manokwari and Fakfak, and in 1902 at Merauke.

In 1907, the Dutch colonial government embarked on a large-scale program to explore their New Guinea territory, run by the armed services. This program involved 800 men, lasted seven years and cost an astounding 5.5 million guilders. Conditions were harsh, and 79 men—one in ten—died in the process. But by 1915 the government possessed a 1:1 million scale map of western New Guinea that was fairly accurate, at least for the coastal regions.

In 1904, even before the large-scale Dutch military expeditions, the army captains Meyes and De Rochemont found what they called the Noord (North) River, in Asmat territory, which led inland toward the snow-covered mountains. In 1907, an expedition under Dr. H. A. Lorentz attempted to reach the mountains via the Noord, but failed.

Lorentz did not give up, and in 1909, an expedition consisting of Lorentz, the Dutch colonial official J.W. van Nouhuys, Lieutenant D. Habbema and his 61 troops, and 82 Dayak carriers from Borneo successfully worked their way up the Noord River to the highlands. Lorentz climbed one of the island's highest peaks, 4,730-meter Mt. Wilhelmina, to the snowline, by his estimate 800 feet below the peak. The Noord River was renamed the Lorentz in honor of his success.

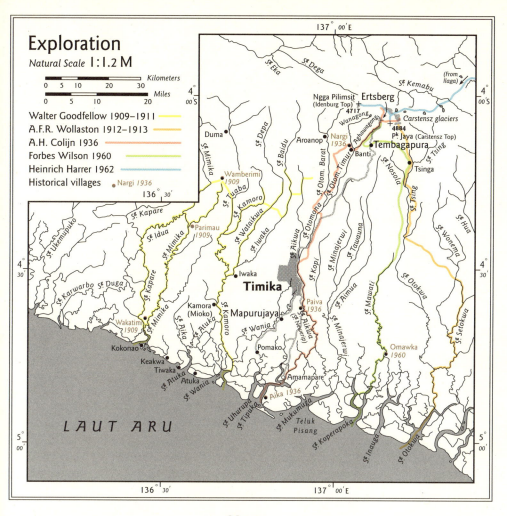

MAP 2.1

When Netherlands New Guinea became part of Indonesia, Mt. Wilhelmina, which had been named for the Dutch queen in 1904, was renamed Gunung Trikora, and the Lorentz River has now—on most maps—reverted to its original Asmat name, the Unir. Also, sometime between 1909 and the 1960s the mountain's ice cap melted, and there is no longer any snow on Trikora.

It had been a point of pride for the colonial government that a Hollander should be the first to reach the snows of New Guinea, and once Lorentz achieved this goal the field was open and the govern-

53

ment granted permission for a scientific expedition of the British Ornithologist's Union to the "Snow Mountains"—the Carstensz glaciers. Only rivers from the south coast offered access to the mountains, which lie almost one hundred kilometers inland. But general geographical knowledge of this region was almost non-existent, and once through the tangle of coastal swamp, the expedition leaders did not know which river offered the easiest route inland.

At the time, the Otakwa was the only river in the area for which inland detail was available, and the leader of the ornithologist's union expedition, naturalist Walter Goodfellow, planned to take this river to the mountains. But the Dutch colonial military and geographical departments in Batavia told Goodfellow the Otakwa would be difficult and suggested the Mimika would make a better route. The British took this advice, and as expedition member Dr. Alexander Frederick R. Wollaston later writes, "This decision, though we little suspected it at the time, effectually put an end to our chance of reaching the Snow Mountains."

The team eventually learned that the Mimika dwindles to a small, jungle-fed stream a full 80 kilometers west of the Carstensz. Whereas the Otakwa, which becomes the Tsing River and then the Nosola River, drains directly from the glaciers.

While the poor ornithologists were struggling up the Mimika, the Otakwa was being further explored by a Dutch military expedition under Captain J. J. van der Bie with the stated purpose of discovering a "convenient" way of crossing New Guinea. The Lieutenant did not get very far, and even today the only convenient way of crossing New Guinea is to fly over it. Van der Bie's team was able to pilot a steamer 28 kilometers up the Otakwa, then journey two days further upriver by launch and another day by canoe. From there, seven days of hacking though the forest brought an advance party to within 25 kilometers of the glaciers. But that was as far as they could go. Although they failed to reach the ice, Van der Bie pioneered the route that would be taken by Dr. Wollaston on his second attempt in 1912.

The accounts of the first British Ornithologists' Union expedition give a good idea of what it was like to travel in this region before the advent of aircraft, roads and accommodations. One of the biggest problems was language, as it was impossible for the expedition leaders to find Papuans who spoke Malay, the language upon which modern Indonesian is based. In his 1912 account of the Goodfellow expedition, Wollaston writes that "the long stretch of the South-west coast from the MacCluer [now Berau] Gulf as far as the Fly River has been quite neglected by Malay-speaking traders, partly on account of the poverty of the country and partly by reason of the shallow sea and frequent storms which make navigation difficult and dangerous, so that the Malay language was of no use to us as a means of talking to the natives."

But the expedition was really defeated by the incessant rain and flooding, the thick forest, and the depredations of disease. "Heavens! how it rained!" Wollaston writes. "Can this forest, with its horrible monotony and impregnability, be equaled by any other in the world?" According to Wollaston's records, during the first year of the expedition, rain fell on 330 days.

"An unprejudiced observer looking in upon us from the outside in the evening might well have wondered what kind of lunatics we were to come to New Guinea," he writes, after months and months still trapped in the lowlands. "If I were to write a true and complete account of the expedition, I should fill many pages with repeated stories of rain and floods, sickness among the coolies and our consequent inaction; but that would be as wearisome to the reader as it was trying to our own patience."

One of Goodfellow's biggest problems was the massive size of his expedition party, and the logistical problems this created. Between the ornithologists, the military escort, and the porters, the total size of the expedition reached 400 men, and supplying this army with food and supplies quickly became an onerous burden. Of the original 400 men, only 15 lasted through the 15-month-long expedition. The rest, stricken by malaria, beri-beri, dysentery or pneumonia, were either sent home or died in New Guinea. Even Goodfellow himself fell to malaria, and was replaced as expedition leader by the team's surveyor, Captain Cecil G. Rawling.

Conditions were particularly bad for the porters—called by the insulting term "coolies" in the accounts of Wollaston and Rawling—who were chiefly young, relatively urbanized men from Ambon, who lacked both the strength and the aptitude for hard jungle work.

The expedition, which arrived in New Guinea in early January of 1910, constructed a base camp at the village of Wakatimi, on the Mimika River just inland of where it is met by the Kapare River. The first stage consisted of ferrying supplies from Wakatimi to Parimau,

The 1936 Colijn expedition, like others before them, hired Kamoro rowers to get team members and supplies from the coast to the beginning of the foothills.

about 40 kilometers inland along the Mimika, near where the foothills begin. At Parimau the expedition met the "Tapiro pygmies," a race Rawling claimed to have discovered. Actually "met" is probably the wrong word, as the hapless men were in fact run down and captured by the team's bearers. The Tapiro were almost certainly Ekagi (also called Me), who still live in the area.

From Parimau, the team made a last-ditch effort upward, eventually reaching just 1,640 meters. Having gained this elevation they were finally able to get a clearer picture of the landscape they had been hacking through. "How different the land looked when seen from above!" Rawling writes. "Where we imagined lay the course of one river, we found another; a hill here, a ravine there, were now exposed to view, though all had been hidden from the level of the plain; and we realized how impossible it is to discover the trend of the rivers in a mountainous country when merely viewing them from low ground."

Though they failed to reach their objective, the British team added a great deal to the western world's knowledge of New Guinea.

Once in the highlands, the Colijn expedition relied on just eight Dayak porters, an improvement over the hundreds of porters used in previous expeditions.

The ornithologists brought back 2,200 bird skins comprising 225 species, many of which were new to science. Mammal skins filled six cases, and specimens of reptiles and insects filled yet more. They brought back a valuable cargo of ethnographic objects, which represents the very first collection of Kamoro art, as well as some useful—although disjointed—information on the life of the Kamoro people. Expedition geologists found a few low-quality coal seams, some indications of petroleum, and traces of tin, copper and iron ore. Most valuable, perhaps, was the 1:250,000 scale map of the region drawn by Rawling and a Mr. E. Marshall, which greatly expanded the geographical knowledge of the region.

After failing to reach their goal, despite 15 months of hard work, Wollaston writes wistfully: "The geographers and the naturalists of the future will live in comfortable ships on the coast, whence they will fly daily into the heart of New Guinea where they will find things undreamt of now. But the time for that is not yet, and in the meantime those who plod on foot do the best they can."

The glaciers were visible to the expedition on one hundred mornings, Wollaston writes, "and any one who has a love of mountains can understand how tantalizing it was, day after day, to see those virgin peaks so comparatively close at hand and yet as unattainable as the mountains of the Moon."

Tantalized he remained, and when the opportunity to lead the second British Ornithologists' Union expedition arose, Wollaston took it. The only European to accompany Wollaston this time was Dr. C. Boden Kloss from the Kuala Lumpur museum in Malaysia, but with 74 porters and a substantial military escort led by Lieutenant R. van de Water, the total number in the expedition rose to 226. Still, this was an improvement, as were the porters, this time Dayaks from Borneo who were familiar with densely forested terrain.

And most important, this time they headed up the Otakwa. After just four months working their way up the Otakwa and the Tsing, Wollaston's party headed into the mountains along a knife-edge ridge between the Nosolanogong and the Beanogong, both tributaries of the Tsing (the suffix -*nogong* means "river" in the local Amungme language). When they arrived in the highlands, a welcoming party of Amungme performed a joyous dance, "forming themselves into a wheel-shaped figure on a small level space, they ran round and round, barking and shouting and waving their bows and arrows," Wollaston writes. This was in marked contrast to earlier encounters further south, in which the team was ambushed "in such determined fashion that firearms had to be used to drive them away."

Once they reached the highlands just south of the glaciers, hiking became easier, but fog and rain were a hindrance. "Every day in those higher regions the clouds descend about 9 a.m., followed soon by rain, so that during the latter part of those days' march we were groping blindly through the fog, and all were drenched to the skin," he writes. "For many days we were never dry night or day, and by reason of the consequent softness of our skins, we were painfully cut by bushes and rocks." Cooking was difficult because of the altitude and wet wood. Rice had to be boiled a very long time.

On the morning of January 30, 1913, when they were so close to the glaciers that their local guides assured them they could reach them and return on the same day, two parties left the camp: one led by Wollaston, and another by Lieutenant van de Water. The two became separated, but both reached the snow, with Van de Water beating Wollaston by half an hour or so. From their 4,300-meter vantage point, the explorers could see Mount Cock's Comb (now known by its Dutch name, Hanekam) to the south, and towering Mt. Carstensz to the west. Two days later, Wollaston made for the snow again, this time reaching, by his estimation, 14,866 feet (4,531 meters), the highest he was to get. The ice-covered slopes of what he had heard the local people call "Ingkipulu"—now usually spelled Ngga Pulu—remained beyond his reach.

"[W]e found our further progress barred on the one hand by precipitous rocks, and on the other by a steep wall of ice, the abrupt termination of the ice-field above. Either of these obstacles might have been overcome by a party of three competent alpine climbers, but for two people to make the attempt in fog and rain would only have been to court disaster," he writes. "[T]o have the prize withheld when it was nearly within our grasp, was almost more than Christian patience could bear." Twenty-three years later, when the Colijn expedition finally reached the summit of the ice fields, Jacques Dozy named the two small south-running glaciers coming off the Carstensz glacier—that part of the snowfield the men had nearly reached—for Wollaston and Van de Water.

By the 1930s, conditions had changed considerably in western New Guinea. Europeans at the few Dutch outposts were adapting to what had been considered an impossibly hostile environment. Expeditions had continued the job of surveying, and the map of the coast was fairly complete. In the south, however, only the tiny outpost of Kokonau, at the mouth of the Mimika River, showed the Dutch flag along the entire 900 kilometers of coastline between Fakfak and Merauke. Although the Merauke station was established at the turn of the century to help stop Marind raids across the border into British New Guinea to collect heads—losing their subjects this way was quite irritating to the British and Australians—the Dutch made no efforts to pacify the equally fierce Asmat living between Merauke and the Mimika areas.

The Dutch did use western New Guinea as a place to put political troublemakers, incarcerating the archipelago's nascent nationalists in the prison camp at Tanah Merah. This death trap, deep inland from Merauke, was guarded on the north by high mountains, and on the south by the Asmat and the Marind, the most warlike and unacculturated cannibal groups in Dutch New Guinea.

Overall, Holland's half of the island was considered by the colo-

The Carstensz snowfields were the Colijn expedition's goal. This is the rock wall between the Carstensz glacier and the Middle Firn. The snowbridge has since melted.

nial government to be an expensive administrative burden. By the late 1920s gold was coming out of Australian New Guinea, and the Dutch began encouraging economic ventures in Netherlands New Guinea. The first takers were oil companies.

Some oil seeps had been noticed in the western tip of the Bird's Head Peninsula, and in 1929–1930, these were explored further by two geologists for Bataafsche Petroleum Maatschappij or BPM, an East Indies subsidiary of the Royal Dutch/Shell group. The program was abandoned because of an economic depression in Holland, but the interest remained.

By the 1930s, BPM was ready to continue its explorations, but by that time the Dutch government did not want to continue issuing individual concessions in New Guinea. The government wanted to grant a single exclusive concession for the entire territory to one entity, so that it would have a politically acceptable excuse for refusing the Japanese companies that had already approached the government seeking oil concessions. Since three companies were already active in the East Indies, they joined forces as the Nederlands Nieuw

The Colijn expedition took advantage of the oil company's old Sikorsky S-38 to scout approaches and drop supplies. Frits J. Wissel is the second pilot from left.

Guinea Petroleum Maatschappij, or NNGPM. This consortium was 40 percent BPM, and through their East Indies subsidiaries, 40 percent Esso (now Exxon) and Mobil, and 20 percent Standard Oil of California (now Chevron) and Texas Company (Texaco).

In 1935, NNGPM received a concession covering 100,000 square kilometers of Dutch New Guinea's coastline and low hills, which was later enlarged to 250,000 square kilometers. To carry out exploration, airfields were constructed at Jefman, a small island near Sorong at the western tip of the Bird's Head, at Serui, on Yapen Island south of Biak, and at Babo, on the south shore of Bintuni Bay.

In 1936 test drilling on the Klamono anticline in the western Bird's Head produced a very encouraging strike. Exploration and test drilling also went ahead on Misool, one of the Raja Ampat Islands off the tip of western New Guinea. Operations were soon interrupted by World War II, but the consortium continued its efforts after the war. Unfortunately, a modest well at Klamono was the only producing result, and by the time the effort was abandoned in 1960 it had lost 575 million Dutch guilders.

EARLY EXPLORATION AT Babo, however, was indirectly responsible for the biggest mineral find in Netherlands New Guinea. In 1936 Jean Jacques Dozy, a young exploration geologist, and one of the world's first photo-geologists, arrived in Babo to work on the aerial mapping of the concession area. Dozy's doctoral researches had taken him to the highest part of the Bergamo Alps in north Italy for mapping work, and his first tour of duty in the Indies was in the sweltering lowlands of Borneo, behind the oil town of Balikpapan. When he arrived in New Guinea, Dozy recalled, the coastal charts were excellent, but maps of the interior were full of blanks. One of his favorite sheets, which he regrets having lost, carried this notation in one of its blank areas: "Here a tax collector has been eaten."

Three DeHaviland Dragon aircraft, based at Jefman, were used in the surveys of the Bird's Head, but the swampy south coast was not considered suitable for the construction of airstrips. Instead, base camps were built at Etna Bay—near Freeport's current exploration base—and at Aika, near the current Kamoro settlement of Yamakupu, just downstream from Amamapare at the mouth of the Tipuka River. These were supplied by an old Sikorsky S-38 amphibious plane, which would also be used for the aerial survey and mapping work. The Sikorsky was rugged and could land equally well in the ocean, in large rivers, and on land, but it took a long time to gain altitude. This failing, Dozy recalled, could be quite frustrating, as by the time it reached the 4,000-meter survey altitude, the coastal plain would often have clouded over.

The manager of the Babo operations was Dr. Anton H. Colijn, a lawyer who also happened to be the son of Holland's prime minister. Colijn was aboard the Sikorsky on one of its first trial flights in New Guinea, to test the aircraft's operating ceiling under local conditions. After an hour of climbing, the plane reached 4,000 meters, and Colijn saw the Carstensz peaks poking above the clouds in the distance. He was a great lover of mountains, and a rabid alpinist, and could not resist having a look. At its maximum working altitude of 5,000 meters, the old plane could just come abreast of the mountains.

When the plane reached the mountains, Colijn made a very important discovery: the Carstensz did not just fall off on their northern slopes. There was a whole system of valleys and glaciers hidden in there. Nor did Colijn neglect to notice a high, flat plain which, he thought immediately to himself, offered the perfect access point to these spectacular glaciers. He called this the "Alpine Meadow." When the team later reached it on foot they found out it was really more of a swamp or bog, but with self-conscious euphemism "meadow" was retained in the name Dozy eventually used: Carstenszweide—pronounced *vay-day*—or "Carstensz meadow."

"You can imagine the excitement Colijn's report caused in the club at Babo that night," Dozy said. "Mt. Carstensz was still 'virgin' and the wildest dreams kept the few alpinists in tropical Babo awake." Dozy and Colijn immediately began laying plans to get there.

Of course, the oil company that employed the two men was not about to let them go off climbing mountains, when their job was to map the region and look for oil. So any expedition would have to be carried out on their own time. The problem was time off. During a standard three- to four-year contract period, employees were allowed only one leave of significant duration, a two-month "Java leave." Also, the cost of the expedition would have to be borne privately, which ruled out elaborate provisions and armies of porters.

The men researched what was then the most successful attempt on Carstensz, Wollaston's second expedition. This expedition, though a model of efficiency for its time, took more than six months, with only a few days at the top. Colijn and Dozy didn't have six months, and they intended to reach the peak. They soon realized that their lack of funds and time might actually constitute a blessing, as they determined that the logistical nightmare of supplying an army of porters and a huge military escort was what stymied Wollaston. The other advantage Colijn and Dozy enjoyed in 1936 was the Sikorsky. "Of course the plane," Dozy later said, "was an essential thing."

By this time a new pilot had arrived at Babo, marine flying officer Frits J. Wissel. Wissel was an expert pilot, a talented writer, and, conveniently, an enthusiastic alpinist. On one of his flights on behalf of the oil company, he became the first outsider to see the Paniai Lakes, which were called the Wissel Lakes in his honor until Indonesia took over administration of the territory. Colijn chose Dozy and Wissel as his companions for the upcoming expedition.

In the months preceding the attempt on the mountains, which was planned for the fall of 1936, Wissel often found time on the return home from survey runs to scout potential routes for the expedition. They already knew that the *weide* would provide the best access to the glaciers, but how to get there? Using oblique aerial pho-

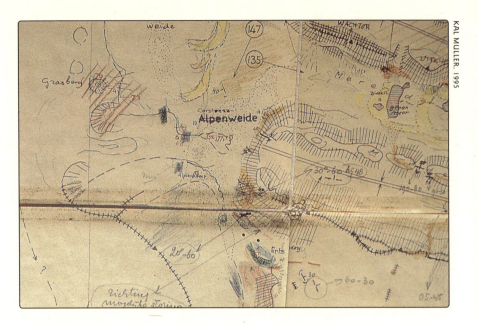

A section of Jacques Dozy's original map of the Carstensz glaciers area. Note that Grasberg, Ertsberg, and the Carstenszweide have been marked and named.

tographs shot from the window of the Sikorsky, Dozy was eventually able to put together a provisional sketch map of the area. The three decided to follow the Aikwa River by boat to the Otomona (where the bridge is now) and from there to follow the ridge top to a confluence of three rivers, at an altitude of about 2,500 meters. They had noticed a settlement there (near present-day Banti Village), which they decided would make a good base camp for the final assault. Because of the altitude and steep terrain, from this village upward the men would have to follow the valleys and gorges.

"The question was: How can we get from the coast to the foothills the quickest? The idea was to follow the Newerip River [now called the Aikwa], cross over, and get up on a crest," Dozy said. "The crest was much easier going. It is far too tough to follow the streams, because then you have to cross all the *cabangs*, little side streams, and that takes much too much time. Incidentally, Freeport later did the same thing when they built their road, although they followed the crest east of ours."

Given the shortage of time and money, the expedition would have

Jean Jacques Dozy, in 1995

to be streamlined, with a minimum of porters. The men figured that if they carried their own backpacks, they could make do with just eight porters. They carried just one big tent, and one tarp to cover it, and everyone squeezed inside at night. The porters were Dayaks who worked for the oil company, and they were as excited as the two Hollanders to reach the *gunung es* ("ice mountain") they had heard so much about.

"It was wonderful to live with the Dayaks," Dozy said. "They were excellent companions, very nice chaps, and of course they had an incredible bush craft. Even the most experienced European doesn't have that nose for the bush. So let's say we 'borrowed' these eight from the company."

The only way to keep the expedition small was to make supply drops using the Sikorsky, putting food and stores at strategic locations just ahead of the advancing party. The big question was how to drop the loads so they would not be broken or lost. Parachutes were the obvious answer, but at the time, Dozy recalled, they were selling on Java for 120 guilders each. Dozy's monthly salary at the time was 500 guilders, so obviously parachutes were out of the question.

Their first thought was to cushion the loads with *ijuk*, a cheap, springy palm fiber used locally to thatch roofs. The men conducted some experiments on the Babo airfield, packing four-gallon gasoline cans in the fiber and dropping them from the airplane. The *ijuk* kept the loads from breaking, but they hit the ground with such speed that they rolled long distances. In the bush, the men reasoned, they would never be able to retrieve their precious supplies.

Then they had an inspiration. "As kids what did we do?" Dozy said. "We took our handkerchief, tied knots in the corners and dropped things with it. Let's try that."

In a nearby hamlet they found a store selling sheets of rough Japanese cotton cloth, four by four meters in size. They bought a couple and tested them. One of these cloths, tied at the corners and carefully folded, would bring a 25-kilogram load safely to the ground. Two tins rigged to one of these makeshift parachutes, if arranged properly, would just fit through the opening in the airplane's fuselage for the aerial camera.

"They opened beautifully and came nicely down," Dozy said. And for 10 guilders each, the cloth was a real bargain. An additional advantage, which the men didn't discover until they were already trekking, was that the cloths, set in a double layer, made nice rain-proof tarps at their bivouacs.

Considering the number in their party, and the distance they had to cover, the team decided they could make do with two drops. The logical sites were the Wa Valley and Carstenszweide.

The expedition started from the oil survey camp Aika (also, in those days, called Amamapare) on the coast. At ten minutes past two o'clock in the morning on October 29, 1936, Colijn, Dozy, and their eight porters packed themselves and their equipment in seven long canoes. Local rowers took them up the Aikwa River to Paiva, a small settlement on the western bank of the river where some Kamoro who worked for the oil company lived. Paiva, about seven kilometers south of where the Timika airport is today, is where the Aikwa begins to spread out in braided rapids. It took fifteen hours to reach Paiva; after another day on the river, the expedition reached the confluence of the Otomona and the Aikwa rivers.

From there on they walked. In two days they reached the foot of the mountains, where they caught up with a group that had been preparing a trail for an upcoming geophysical exploration party. Then Colijn and Dozy headed along the ridge crest between the Aikwa and the Otomona rivers, in many places having to cut their own trail. They worked steadily along the ridge, crossed the Western Otomona where it is met by the Simpang River, and continued along the crest between the Western and Eastern Otomona rivers.

At night they made camp, eating a simple meal of rice, beans and dried fish. The hardest part, Dozy later said, was knowing where to come down. As they reached the higher elevations, the cloud cover often was such that their aerial photos and Dozy's sketch map were not much help. Finally, they left the ridge top, coming down into a settled valley near the hamlet of Nargi (near what is now Opitawak, just west of Banti), and made their first contact with the Amungme.

"This was rather a critical moment, because we had told the authorities we needed no military people to cover us," Dozy said. "Of course the villagers knew we were coming, so they were all standing with bows and arrows on a little plateau as we came out of the bush. What should one do? Suddenly Colijn, shouting and singing, rushed to them, and that was very, very funny. Before we knew it, the tension was broken."

Since the team was right on schedule, Wissel soon flew over.

Members of the Colijn expedition resting on the way to their goal. From left: pilot Frits J. Wissel, expedition leader Dr. Anton H. Colijn, and Jean Jacques Dozy.

They signaled him that everything was all right, so he went straight back to Aika and fetched the supplies. Colijn and Dozy had brought some steel knives for barter, and told the men of Nargi that they would be given a knife for every one of the dropped cases they retrieved. "In no time we had all our supplies," Dozy said. "The food drop went very well, and we only lost one box."

The ground team established a base camp near the Amungme settlement. Wissel made his second drop on the meadow, took the plane back, and hustled upriver to join the expedition. While waiting for Wissel to catch up, Colijn organized the camp and Dozy, who had the most bush experience of the two, scouted trails to the alpine meadow. Dozy found the Amungme to be very hospitable. Scouting trails with one of the Dayak porters one day, he came upon two women working in the very highest garden plot.

"When they saw us," Dozy recalled, "they said 'It's too cold where you're heading, you shouldn't go there!' We said, 'We're going there anyway.' Seeming concerned, they suddenly dug up some potatoes for us, which was awfully nice of them."

Members of the Colijn expedition on the summit of Ngga Pulu, then the highest point between the Himalayas and the Andes. From left: Dozy, Wissel and Colijn.

Once the men ran out of Amungme garden trails and hunting trails, the hard work of cutting their own began. The gorge cut by the Aghawagong (just below where the mill is now) is where Dozy saw the last trace of human use. From that point on, it took weeks to cut tracks that would take just days to walk.

"That was tough cutting," Dozy said. "Only one man could cut at a time, so I took turns with the Dayaks and it was *awfully* tough bush. Not high, but dense and tough. You worked the whole day, soaking with sweat, and in the end you walked home and were back in camp in 20 minutes. And you had to feel your way. We had that map and aerial photos, so we had something to orient us, but in the bush you don't see much."

Then came the toughest part, the climb up the sheer mountain wall to the alpine meadow. It was Dozy's task to find a way up the cliff. Looking for the easiest way up while being drenched by freezing waterfalls and rain was a miserable experience. The job took three days. At one point, blocked by a sheer rock face, Dozy and his men had to cut down a tree and fix it in place against the rock with a rope.

"That was the toughest part," Dozy said. "Now that gully is filled up with debris, but at the time it was just a gorge with cascades. It was cut off by ledges and limestone ridges that crossed it, and you had to try to find your way around—or over—those ridges. And the rain. It rained like hell and after a while the Dayaks said 'We can't walk any more!'"

This initial scouting work made the climb fairly easy for the rest of the expedition. The Dayaks even cut a rough ladder to replace the tree trunk. Of course, these were very experienced men, and even today this climb is not recommended for anyone but expert climbers. It's a lot easier to take the tram.

On Monday, November 23, 1936, Dozy finally broke through to the Carstenszweide. His journal entry for that day details this effort:

> I leave Bivouac 11 at 5:50 a.m. with Tadong and Besi. The weather does not look good and it is very cold. In a small ravine where we erected a few poles yesterday to manage a rock wall, an icy wind is blowing down on us. At 7:30 we are at the end of the trail cut the previous day, and we are happy that we can start cutting our way up so early.
>
> My hands tingle with cold and our feet are icy lumps. The shrubs are very dense and wet. Progress seems hopelessly slow. My feelings of frustration are swept away, however, when I climb a low tree along our trail and fail to see any serious obstacle ahead of us any more. Shall we at last reach our goal, the alpine meadow, today? After a certain distance we reach a stream which we follow for a time. Then the mood is suddenly on me: we must break through to the alpine zone today. I jump ahead through the cold waters of the stream and over boulders; we climb up against the rocky faces below waterfalls. At least we make headway. Around gorges too narrow to manage directly we still have to cut a trail. To our joy we come upon some banks along the creek covered only with low brush and grass, which makes walking much easier. Soon gentler slopes follow where shrub is getting sparser. We are now on the east bank of the stream and reach a comparatively rolling, peat-like surface. It started raining in the meantime. The glistening black rock walls to our right show large patches of the green colors of malachite. The tones are reddish, green, white and gray. The bushes show bright red flowers. In the foggy atmosphere it all creates a queer effect.
>
> Through a shrubby vegetation we follow the foot of the rock wall till at last a few vague contours loom up in the gray light of rain and mist, showing in yellow-gray tints the flat bottom of Carstensz meadow.
>
> —*translated from the Dutch by J.J. Dozy*

Tadong and Besi had the worst of it, as Borneo doesn't have mountains this high and they had never experienced such cold. At a bit before one o'clock, the team headed back down from the "meadow," which was a marsh consisting of great tussocks of grass. They were all completely soaked, and Besi nearly gave up from the cold. Dozy had to keep reminding him of the warm fire back in Bivouac II. As they neared the camp, they shouted four times, the sign that they had made it to the *weide*. By the time they arrived, there was hot oatmeal waiting.

Wissel soon caught up with his partners, and the expedition moved camp to the Carstenszweide on November 26th. Two days later, Ertsberg was first surveyed. The language of Dozy's journal entry for that day is almost disappointingly dry:

> On Saturday 28th the geological section exposed on the east side of the Carstensz meadow was visited and surveyed. It confirmed a synclinical feature of Upper Tertiary limestones with foraminafera allocated to an Upper Tertiary age. Southward the limestone becomes sugary and even coarsely crystalline.... At the end of the traverse we came to the ore body in which chalcopyrite, malachite, lazulite and bornite were recognized, and noted that the body might have a width on the order of 100 meters. Several samples were collected.

Dozy dubbed this ore body "Ertsberg," Dutch for "ore mountain," which he penciled in below a sketch he had made of it two days previously (see photograph page 72).

No geologist would have missed Ertsberg, it was just too obvious. "I knew in a blink of an eye what this was about," Dozy said. "It was hard to miss, with all the green and blue spots. The copper was obvious." Grasberg, too, was duly noted in Dozy's report, published in 1939, as were all the other main features in the region. In retrospect, it seems odd that this report generated so little interest. But even when he first saw it, Dozy said, "I realized that no one could do anything with it. There were no roads, no harbors, no factories. It was just like a mountain of gold on the moon." Dozy's superiors at BPM read the report, and shrugged their shoulders. "What can you do with it?" they said.

And of course, Dozy, Colijn and Wissel were on their way to the snow. Within a week of reaching the *weide*, they had established themselves at the Glacier Bivouac, in the Yellow Valley near the Carstensz glacier. At two o'clock in the morning on December 5th, the men set out under the light of the moon for the highest point in the Carstensz mountains, and in fact, the highest point between the Himalayas and the Andes: Ngga Pulu. They reached it in just five hours. Dozy's barometer read 5,030 meters at the summit of Ngga

Jacques Dozy's first sketch of the Carstenszweide, made on November 26, 1936. The next day he penciled in a name for the drawing's central feature: 'Ertsberg.'

Pulu, but correcting it for error shows that they actually reached 4,906 meters. The glacier has melted since, and Ngga Pulu is now at least 45 meters lower. But in 1936, the men stood 22 meters higher than Puncak Jaya, currently the pride of the Sudirman Range.

WORLD WAR II interrupted mineral exploration in the area, and Dozy's report, published in a very small edition by the University of Leiden, languished in a few libraries while the German army overran Holland. It was to resurface only in 1959. A Dutch mining company called Oost Borneo Maatschappij (OBM), the East Borneo Company, had had coal and nickel mines in Borneo and Sulawesi, which they lost after Indonesian independence. They still had 70,000 guilders left and were looking for nickel prospects in Dutch New Guinea. Jan van Gruisen, a mining engineer, conducted a literature search and came upon Dozy's report. He didn't think much of it, but he took the precaution—which was in any case quite inexpensive—of applying to the Dutch government for a concession on a ten-kilometer by ten-kilometer square of land centered on Dozy's Ertsberg. OBM

The same view as Dozy's sketch at left. Looking south across the Carstenszweide, towards the Aghawagong Valley, with the black mass of Ertsberg at its head.

received the permit, but the company didn't really have the money to explore, much less develop the area.

Van Gruisen had a mining friend from the United States whom he had known for years. They had even worked together on a couple of nickel projects, including one on Sulawesi. This friend was Forbes Wilson, then chief of exploration for Freeport Sulphur.

Wilson happened to be in Europe on a business trip in 1959, just a few weeks after Van Gruisen found a copy of Dozy's report. Van Gruisen handed Wilson some pages from the report and asked him what he thought. "My reaction was immediate," Wilson later wrote. "I was so excited I could feel the hairs rising on the back of my neck." It wasn't just Ertsberg, but the whole geology of the region described in the report that attracted Wilson's attention. "The Ertsberg, it seemed likely, might be merely the surface outcropping of even larger copper deposits underground."

The report Van Gruisen had rediscovered was the 1939 "Geological Results of the Carstensz Expedition 1936" published in Leiden. In summarizing his analysis of Dozy's ore samples, Dr. Ir. C. Schouten of Delft writes:

> The ore-samples can be called contact-metasomatic gold-bearing copper ores. In so far as we may conclude from a sample which is by no means representative—a few loose clumps of ore from the surface were the only available—the copper-content appears to be high (0–40%), while the gold-content might be rather considerable (0–15 gr. per ton). ...
> Even if the ore should prove to occur in large quantities, an economical exploitation will hardly be possible owing to the remote position of the locality. The costs of transport will be too high. Moreover the metallurgy of the ore would not be simple, as can be inferred from the mineral assemblage and from the data obtained by the microscopical examination.

Wilson wasted no time in finding Dozy at The Hague, and the Dutch geologist confirmed that there was a large deposit of seemingly high-grade copper ore lying exposed in the middle of New Guinea. Wilson was a man possessed. "I must see the Ertsberg for myself or die trying," he writes. Van Gruisen said that OBM, which lacked the money to finance an exploration effort on its own, would consider a joint venture with Freeport. Wilson cabled New York—then Freeport headquarters—and received authorization to spend $120,000 to evaluate and sample this deposit. For the then 50-year-old man, it was going to be the toughest trek of his life.

In the late 1950s the sum of $120,000 would compare to maybe $1 million today. It makes us think, in today's corporate America, could a geologist or mining engineer get this kind of backing to explore an ore body in such a remote and unknown part of the world? I doubt it. It is a tribute to Forbes' persuasiveness, and to the willingness of the Freeport managers at the time to accept risk, that he was able to mount this expedition.

Forbes Wilson's struggle to the Ertsberg forms the bulk of his very readable book, *The Conquest of Copper Mountain*. He gave up a 30-year smoking habit, received inoculations for just about every disease known to man, and outfitted himself for both jungle and alpine conditions. An advance team was sent to Biak, and then to the south coast, and in April 1960 Wilson, accompanied by John Bowencamp, a friend and Freeport mining engineer, left for New Guinea.

The advance team waiting for Wilson and Bowencamp consisted of Delos Flint, Freeport's chief field geologist and a personal friend of Wilson; Augustinius "Gus" Wintraecken, a Dutch engineer who spoke Indonesian fluently and had extensive experience in Indonesia; Jan Ruygrok, a young Dutch ex-Marine and botanist; and Vincent "Bob" Croyn, an Ambonese-French police officer. These four had constructed a base camp at Omawka, a village on the Mawati River where it is met by the Omawka River, near today's Fanamo and Omawita villages.

Poring over U.S. Army aerial photographs of the region, the expedition leaders had earlier thought they could quite painlessly approach Ertsberg from the Paniai Lakes in the west. Van Gruisen from OBM flew out to New Guinea to meet the advance party, and they chartered a Grumman Mallard to fly over the proposed route. Unfortunately, the angle of the U.S. Air Force photographs disguised the true character of the Ekabu Plateau that lies between Paniai and Ertsberg. Instead of smooth rock slopes, the expedition would have to cross rugged limestone, shot through with vertical canyons and crevasses, often several hundred meters deep. So, like the others that preceded them, the Freeport–OBM expedition would have to contend with the swamps and steep mountains of the south.

Although OBM jointly sponsored the expedition, the company never joined in developing Ertsberg. A subsidiary of the East Borneo Company, however, retained a 5 percent interest in Freeport Indonesia's operations until it was bought out by Freeport-McMoRan in 1988.

Wilson expedition member Jan Ruygrok, a former Dutch marine, generates power for the radio so the team could order supply drops, and issue progress reports.

The advance party first headquartered in Sorong, on the Bird's Head, where they chartered the *Seremoek*, an oil company ship. They also hired some local Sansaporese, who were mountain people, although, as Del Flint later said, "there ain't no mountains in the western end of New Guinea like the Carstensz." Though not quite up to the rigors of the Carstensz, the Sansaporese were very good workers, Flint recalled.

Crucial to the success of this expedition was "Moses" Tembak Kilangin, an Amungme from the Tsing Valley. While Van Gruisen was in New Guinea helping scout routes to the highlands, he stopped in Kokonao to discuss possible ways inland from the south with Father J.P. Koot of the Catholic mission there. There he was introduced to Kilangin, who was Catholic-educated and could speak four languages. He proved essential in organizing porters and settling disputes in the highlands. Now more than 70 years old and living in Harapan, Kilangin still receives a modest stipend from Freeport.

The reconnaissance team first followed the Mawati River, then crossed over to the Tsing River, following the ridge line west of the

The Wilson expedition required 14 canoes and 44 Kamoro paddlers to haul the men and their supplies up the Mawati River to the beginning of the highlands.

Tsing. This was Wollaston's 1912 trail. When they reached the highland villages in the Tsing Valley, they headed west to the Wa Valley, from there following Colijn's 1936 route. The Freeport–OBM expedition established a base camp about three kilometers northeast of the Dozy expedition's main staging camp, where Utekini is today. They cut a trail through the thick vegetation north along the Aghawagong Valley and, like Dozy, fought to cut a trail up to the Carstenszweide.

Flint quickly came to admire the Amungme he met working on the trail to Ertsberg. "These guys have a hell of lot on the ball," he said. "They could start a fire in sopping rain, and they could do it faster with a piece of rattan using friction than we could with kerosene. They lived in conditions we couldn't live in, and they did it using their heads."

On the distinctive sartorial feature of the mountain Irianese, Flint said, "The first time I saw this guy standing there—even though I had seen pictures of these penis *kokers* [old Dutch term for penis sheaths]—I felt just *absolutely* outclassed. I thought, My God, how can anything compete with *that*?"

Flint reached the Ertsberg a few days before Wilson, and couldn't resist the urge to climb up on top. "I got to the top and was able to see this glacially polished expanse of magnetite with big patches of chalcopyrite and it was a feeling like, My God! What a deposit! It wasn't an intrusion, but a metamorphic mass."

Ertsberg was certainly an inspiration to a geologist. Of his first sight of Ertsberg, Wilson writes in *The Conquest of Copper Mountain*:

> It was much larger than I had imagined in numerous prior attempts to visualize it. I stared at it for several minutes before slowly, almost reverently, approaching its base. I had known about the Ertsberg for less than a year, but I felt like I had been waiting for this moment all my life. With my geology pick, I struck several of the boulders on the ground that had broken off the main mass. When the thin black outside layer—the product of oxidation—was chipped away, I saw on every piece the gleaming golden color of chalcopyrite, a sulfide of iron and copper. Everything Dozy had told me at our meeting at The Hague seemed confirmed.

Based on the 300 kilograms of surface chips the expedition brought back, the Ertsberg turned out to be 40–50 percent iron, mainly in the form of magnetite and iron oxide, and 3.5 percent copper, chiefly chalcopyrite and bornite, both sulfides of iron and copper. This was the largest and richest copper deposit ever found above ground. The assay data confirmed the expedition's estimates of "13 acres," Wilson's special code for 13 million tons of ore. Below ground, it was further estimated, lay 14 million tons of ore for each 100 meters of depth, for a total of perhaps 50 million tons of ore.

But surface samples and educated guesses, even by experts, are not enough for a company to spend the many millions of dollars necessary to extract and process the minerals. Especially when the deposit is, as Dozy said, "like a mountain of gold on the moon." There is only one way to reassure investors and obtain loans. Diamond drills must be brought in to pull out core samples from deep within the ore body to convincingly establish the size, shape and contents of the rock. Interesting chips from the surface won't do. Ertsberg needed to be drilled.

CHAPTER THREE

Building a Mine

Turbine helicopters—Irian Barat—New leadership in Indonesia—A Contract of Work—Drilling Ertsberg—Darnell Ridge—Boardroom doubts—An impossible road—The first concentrate shipment—Tembagapura

To PROCEED WITH the Ertsberg project, Freeport needed to systematically drill and evaluate the deposit. But in 1960, it took 17 grueling days for Forbes Wilson and his expedition just to get themselves to the mountain, and there was simply no way to haul several-ton drill rigs using dugout canoes and local porters. Helicopters were the only answer. In the very early 1960s, the most powerful helicopters available used piston engines, and could lift no more than one passenger and a 120-kilogram payload to Ertsberg's 3,600-meter elevation. With this limitation, just getting one drill rig and a crew to run it to Ertsberg would take months, and the only rig that could be lifted by helicopter would have a very limited depth capacity, and do us little good.

By the early 1960s, however, turbine-engined helicopters were already on the drawing boards, and in the heady days of the space program and the Vietnam War, aviation technology was progressing rapidly. Within a few years, more powerful turbine-engined helicopters did become available, and these were capable of lifting 450 kilograms to the elevation of Ertsberg. This was barely adequate for the drill rigs we needed, but barely was good enough. Helicopter technology was not our problem. What almost killed the Ertsberg project was the changing political situation in western New Guinea.

In 1960, Ertsberg was in Netherlands New Guinea; in 1963, the deposit was in Irian Barat. When Indonesia won its independence from Holland in 1949, the western half of New Guinea was the only part of the Dutch East Indies retained by the Netherlands. The Dutch argued that New Guinea had no cultural or geographical relationship to the rest of the archipelago, and thus should not rationally become

Helicopters were assembled in Timuka, the coastal base camp for the Ertsberg drilling program. At the highest tides, five feet of water covered this grassy plain.

part of the new Indonesian nation. Indonesia accepted these terms only under protest.

Throughout the 1950s, President Sukarno appealed to the United Nations to force the Dutch to relent, and allow Netherlands New Guinea to be incorporated into Indonesia. In Holland, however, right-wing groups, indignant at the loss of the Indies, and Christian groups, intent on continuing their evangelical work in New Guinea, fought effectively against giving up this last vestige of the Dutch colony. In late 1957, having lost his case on the floor of the United Nations, Sukarno took action at home: Dutch citizens were expelled from Indonesian soil, and most Dutch assets were seized. In 1961, he landed paratroopers at scattered locations in West New Guinea to expel the Dutch by force.

The battle for New Guinea riveted the young nation's attention, and the entire country united behind Sukarno. The military confrontation, led by General Suharto, ended in a stalemate, but Indonesia won the war on the political front. Holland was not prepared for another unpopular colonial war. The United States, anxious to avoid

Timuka, now sometimes called Timika Pantai, in 1995. Though now overgrown, the old Japanese airstrip is still visible inland and parallel to the village.

driving Sukarno even further towards the Soviet Union, pressured the Dutch to acquiesce. In August of 1962, Holland accepted a plan mediated by Ellsworth Bunker of the United Nations to transfer the territory first to the United Nations, then, in 1963, to Indonesia. A referendum was scheduled for 1969 in which the people of western New Guinea would vote their allegiance.

In 1963, now in Indonesian hands, western New Guinea was renamed "Irian Barat," or West Irian. This resolution of the conflict over claims to the territory might have been good for Freeport's Ertsberg project, except that Indonesia under Sukarno was moving further to the left, and growing ever more hostile to foreign business. By 1963, most foreign-owned operations in Indonesia had been expropriated, Indonesia had withdrawn from the United Nations, and the nation's president was cultivating close ties with Communist China and the Indonesian Communist Party, the PKI. No bank or other sane investor was going to back our project under these conditions.

But in 1965 the country's political situation changed dramatically. Sukarno's power crumbled after a failed coup attempt, and General

Suharto, who speedily put down the insurrection, rose to power. By early 1967 Suharto had been formally named acting president—he became president a year later—and Sukarno was under house arrest. Suharto, in stark contrast to his predecessor, was strongly anti-communist and very pragmatic in matters of economics. He began his presidency by courting foreign investment.

By 1966, the stage was set for Freeport to continue with its Ertsberg venture. The political disruptions of the previous five years had led the company to direct its attention elsewhere, but the project had not been forgotten. The chairman of Freeport at the time was Langbourne Williams, who heard some encouraging news about Indonesia from two executives of Texaco, a company that had been able to retain its large investments in Indonesia even during the height of Sukarno's period of expropriation. Texaco was successful in large part because of its Indonesian manager, Julius Tahija, a savvy and well-connected businessman. Tahija, who is Ambonese, was not part of the Javanese power structure, but he had served in the military and enjoyed close ties with the government.

Tahija arranged a meeting in Amsterdam between Freeport executives and General Ibnu Sutowo, Indonesia's minister of mines and petroleum. Although Freeport had tried to keep its expedition quiet, Sutowo was quite familiar with the expedition and the promising preliminary results from Ertsberg. He told the company that the Japanese, in particular, had approached him about taking on the development of Ertsberg, but since he felt that we had the best chance of bringing the mine to production quickly, he encouraged Freeport to start up again where it had left off.*

Freeport president Robert Hills and Forbes Wilson asked Tahija whom he would recommend to help the company negotiate a contract with the Indonesian government, and he suggested Ali Budiarjo. Budiarjo, who would become crucial to the success of the project, was a member of Indonesia's inner circle, having served in the 1950s as secretary-general of defense and director of national development. He fell out with the Sukarno government in the late 1950s, and during that time took a masters degree in industrial management at the

---

*Years later I heard from a friend in the department of mines that in 1966 a Japanese group was actively being courted by the department to develop Ertsberg. The Japanese organized an expedition to evaluate the deposit, but, as my friend said, "they couldn't even find it." The department quickly lost interest in them.

Massachusetts Institute of Technology. By 1966 he had set up a private law firm, and Freeport became his first foreign client. In 1974, taking over from Forbes Wilson, he became president of P.T. Freeport Indonesia.

When the Freeport executives arrived in Jakarta in June 1966 to begin contract negotiations, Budiarjo's first task was to educate them on the finer points of Javanese business etiquette.

"The Americans were ignorant of their political and psychological differences with Indonesians," Budiarjo said. "Indonesians are not outspoken. They are taught not to question—much less oppose—the opinions of their elders or superiors. I had to explain to the Americans how to draw opinions out of the Indonesians. Also, I told them that they could behave like Americans, but like *polite* Americans, especially with the authorities."

At first, the Indonesian ministry of mines offered Freeport a "production sharing" type of agreement, based on guidelines developed under Sukarno for foreign oil companies. The provisions of this agreement were palatable only to some quick return oil operations, and totally unsuited to the heavy investment and long lead times of the copper mining industry. The committee understood these problems, but they simply didn't have the authority to create another framework. They asked Freeport to prepare a draft contract that would be acceptable to the company. Freeport legal counsel Bob Duke prepared a document based on a "contract of work" type of agreement, which had been used in Indonesia before production sharing had come along. Put simply, a contract of work falls somewhere between the older colonial "concession" arrangements, in which the foreign contractor gets full ownership of the minerals and even the land, and production sharing, in which the host country immediately gets title to all equipment and improvements, and after a short period owns the entire operation.

Behind the scenes, Budiarjo labored quietly. This was not simply out of loyalty to his employer, but also because he believed American investment would be good for Indonesia. By January of 1967, Budiarjo was further armed with the new Foreign Investment Law, which completely reversed the leftist and exclusionist economic policies of the Sukarno regime.

"I believed that the future of Indonesia was only possible with investment, domestic or foreign," Budiarjo said. "Domestic capital at that time was too little to count. Since Freeport was the first foreign capital investment under the new law, it *had* to be successful, as an example for other potential foreign investors."

The Freeport negotiating team included Nils Kindwall, one of the "Founding Fathers" of Freeport Indonesia. The story of these negotiations, Kindwall says, would make a book in themselves:

"When we first arrived in Jakarta in early 1966, Suharto had just taken over, there was no Foreign Investment Law, we had competition for the Ertsberg, the infrastructure was a mess, there was only one hotel in Jakarta, the economy was in tatters, the legal basis for an agreement was vague, and Irian Jaya's national status was still subject to a forthcoming act of self-determination. We were the first to be willing to take a risk with the newly established government. It may seem obvious now, but at that time it seemed crazy to some.

"The small problems were numerous. There was one typewriter in the hotel, which we rented by the hour and on which we did our own typing. Our telexes—no faxes then—back to the U.S. could be read by anybody. The 'characters' around at the time and the rundown airport made it all seem like an Ian Fleming novel. We rode in a 15-year-old Chevy with no instruments, bad tires, and certainly no air conditioning."

The negotiations were nearing successful completion when the Indonesian authorities made a series of tax demands so onerous that the Ertsberg project would have become economically impossible. "The sharp and sudden swings in our fortunes," Wilson writes, "were beginning to make me feel a little manic-depressive."

Eventually, everything was ironed out. The goodwill established between Freeport representatives and the Indonesian government during these negotiations would help both parties through several difficult episodes in the years to come. On April 5, 1967, as the television cameras rolled, Saleh Bratanta, the new minister of mines for hard minerals, and Freeport representatives signed a thirty-year contract of work to develop the Ertsberg mine. This was the first contract signed by Indonesia under the new Foreign Investment Law.

Ron Grossman, in charge of taxes and other financial issues for Freeport, recalls that the initial contract was held together less by its language than by the two parties' shared sense of fairness and willingness to work together. During initial periods of financial hardship for either the company or the government, the parties would meet

DEL FLINT, 1967

The drill camp in the swampy Carstenszweide, at the base of Ertsberg. Though built on Ertsberg's talus slope, the buildings still required foundations of two-by-twelves.

and work out any problems arising from their agreement. For example, in the first year of operation, while the company still enjoyed a tax holiday, copper reached $1.40 a pound. Freeport could have made a lot of money, and the government would have received nothing.

"They were upset," Grossman said. "They came back to us and said, 'We've got to get rid of this provision.' And we did. We adjusted that out, and they agreed to change some things that were a problem for us. Nothing was accomplished unilaterally, it was give and take. It was a very, very good relationship."

Later in the 1970s, copper prices plummeted, and the company faced serious financial difficulties. The company's lenders couldn't help, and the customers said: "Shut it down. There's plenty of copper. Mothball it." But the government of Indonesia asked us, "What can we do to help?" Without this willingness, Freeport Indonesia would never have made it. During the first seven years of mining, Ertsberg had earned the company's shareholders only $15 million, on an investment of almost $300 million—not a good return by any measure.

We negotiated an arrangement with the government to get us

through a two-year period when we were under serious financial stress. Under our contract, we had to pay what was called a "minimum tax," regardless of our profitability. The government agreed to defer the payment of all taxes and royalties during this two-year period, interest-free, which allowed us to maintain cash flow during this tough time. You could never have convinced the U.S. government to do that.

As the first contract of work under the new Indonesian law, the agreement broke new ground. Freeport's tax department helped the Indonesian government develop its system for taxing expatriates, something with which it had no experience, even opening its salary books so the tax office could get a general idea of the level of expatriate pay. The good relationship between Freeport and the Indonesian government also established the notion of the primacy of a contract, that the terms of a contract are not affected by future changes in, for example, the Indonesian tax code.

During the 1970s and early 1980s, the director general of taxes was western-educated, and he took a great deal of pride in Indonesia's obligation to honor our contract of work—even though there were stipulations in the contract he wished they hadn't agreed to. But he always felt that the agreement was bilateral, and that both sides should honor it. This is what we admire so much about the Indonesian government—it respects the objectives of the contract.

WITHIN A MONTH of signing the contract of work, the company was preparing for the next step, bringing in drill rigs to obtain core samples from deep within Ertsberg. As a base, Wilson picked Timika (now called Timika Pantai or Timuka), a small village west of Amamapare at the mouth of the Tipuka River. During World War II, the Japanese had built a fighter strip there, and the Dutch had on occasion landed DC-3s on it. In Darwin, Australia—Freeport's logistics center until a typhoon knocked it out in 1974—the company acquired generators, a bulldozer and a truck, and loaded them onto a converted World War II landing craft, the LCT *Turtle*.

Ted Fitzgerald—alias Ever Ready Ted, for his readiness to fight—brought the first load across the Arafura Sea in May 1967. As he neared the coast of Irian, an Indonesian navy patrol boat intercepted the *Turtle* and fired a shot across its bow. Everyone was arrested and escorted to Kokonau, charged with illegally invading territorial waters. The Australians spoke no Indonesian, and the Indonesians no English, so the problem wasn't resolved until the navy patrol's captain learned by radio that his superiors had forgotten to tell him Fitzgerald's ship had clearance.

Wilson chose his old friend Balfour Darnell to head Phase II, the drilling program. The two men had worked together for years in Colombia, and Darnell was Wilson's best man at his wedding. "Forbes and I were about as different as you can find," Darnell says. "He was Ivy League and I was a roughneck, but we hit it off fine." When Wilson called him up in 1967, Darnell—then 60 years old—had retired. "He wanted me to go to New Guinea," Darnell said. "I had no idea where the hell I was going." H.I. "Pete" Peterson, vice president in charge of Freeport's Australia office, and his second-in-command, Reg Barden, were also crucial to the Phase II operations.

In August, Darnell was in Darwin, Australia, recruiting help. He found 18 husky Australians "off the beach," and loaded them onto the *Heramia Star*. They reached Timuka, and since the boat couldn't get any closer than 150 meters from the shore because of the shallow water, they shuttled themselves, their tents and their gear to shore in a small aluminum boat. Fitzgerald was a few days behind them in the LCT, with a load of pre-fabricated buildings. The first task was to clear some land back from the beach and set up a camp.

"It was pretty swampy," Darnell recalled. "And we got there at the beginning of the monsoon. Well, I had heard that the monsoon was wet, but talk about a lot of rain and a lot of wind! We were in tents and pretty miserable for about a month."

While clearing the land for the camp, Darnell learned from the locals about what they called the "king-of-tides." This is an occasional freak tide four or five feet higher than normal. He decided to put all the buildings on five-foot posts just in case. And sure enough, the tide came. Darnell even had the foresight to raise the septic tank, and during the king-of-tides it stayed above water.

Wilson wanted the advance team to renovate the old Japanese airstrip, so that supplies could be brought in by plane, avoiding the trip across the rough and unpredictable Arafura Sea. Unfortunately, after spending a lot of time and energy cleaning up the strip, Darnell found that at the highest tides it was covered by ten or twelve inches of water. This ruled out airplanes altogether.

Three helicopter pilots and two mechanics had arrived with Dar-

Ertsberg, the largest above-ground deposit of ore the world has seen. In 1968 the monolith

...s above the drill camp, visible here at its base on the Carstenszweide.

nell, and once the camp was set up, a load of supplies, including two disassembled and crated Bell 204B helicopters, arrived by cargo ship. Because of a labor dispute on the boat, which could get only within four kilometers of the shore because of the shallow water, the men had to unload the cargo onto the LCT themselves, a tedious process that took days. They pulled the crated helicopters up onto the grassy plain behind the beach, and began assembling the machines. To the Kamoro living in the area, who had no idea what this machine was, the process was fascinating. A gallery of entranced observers—at least two hundred strong by the end—came out every day to watch the aircraft being assembled.

On September 20, 1967, when chief pilot Don La Freniere climbed in to test the first helicopter and closed the door, the crowd surged forward in panic, uncertain of what would happened. La Freniere opened the door again to get them to move back so he could start the engine. When the engine fired, the screaming turbine and rotor wash drove the crowd back into the mangroves.

"The next order of business was to find the damn Ertsberg," Darnell said. He had two old maps. An almost useless 1:1 million scale map of the whole south coast, and a fuzzy blueprint of a 1:100,000 scale World War II Dutch military map of the region. He also had pictures of Ertsberg taken by Wilson, and Dutch army photographs of the area around the Carstensz glaciers. "I couldn't see the Ertsberg, but it looked like it had to be somewhere in there," he said. Darnell worked chiefly from a compass bearing—30 degrees east of magnetic north, from the very visible erosion pattern at the headwaters of the Kopi River.

"Most of the time we were aborting flights," Darnell said. "That's the monsoon weather over there. You get up at five o'clock in the morning, and wait for the dawn to see what was going to happen. Soon as we'd get a hole, we'd go up through. If we could see the range, we would go towards it. We flew VFR, visual flight rules—no instrument flights with a helicopter. It took us ten days to finally find the Ertsberg."

On the day they reached it, they were at about 2,000 meters when the clouds started closing in over the foothills. They had a safety pad on a sandbar on the Aikwa River, not too far from where the airport is located today. So they sat down on the pad to wait out the weather. When it broke up, they took the machine straight to 3,600 meters, above the clouds, and climbed slowly to 4,300 meters, looking down through breaks in the cloud cover.

"We looked down in the slot and there was the Ertsberg. Don [La Freniere] just flattened the pitch of the rotors, and we went straight on down and flew right over the top, and went back up into the valley behind it to find a place to land."

Darnell wanted to shoot some pictures and to get some idea of where a camp could be built. He jumped out as soon as the aircraft touched down, and started running across the Carstenszweide. He was soon up to his knees in what he calls "muskeg," the sphagnum swamps of the northern United States and Canada. In his excitement, it didn't at first register that he was standing in a swamp.

"My first view of the Ertsberg was when I got stuck in the muskeg," he said. "I was fascinated from then on. I'd never seen anything like it. There was thirteen-and-a-half million tons of ore, sticking up there like a sore thumb. First time I ever saw a mine sticking up in the air."

Darnell snapped out of his trance when it suddenly occurred to him that the helicopter was wound up to full power. He looked over and saw that La Freniere had the machine at full throttle trying to pull it out of the muck. The pilot waved Darnell over, but he put him off until he could get some shots of a potential camp site near Ertsberg. La Freniere, anxious to get out of there before they were stuck for good, brought the helicopter over to Darnell, who climbed on the skids and up into the cabin.

Once they found the site, they began to drop tents and equipment. It took two weeks to move eight men and their facilities to the Ertsberg camp. Because the helicopters were new models, the mechanics had cautioned the pilots not to shut down the turbines above 3,000 meters, and engines idled nervously while the men unloaded. Then one morning the helicopter dropped down and Darnell noticed eight wooden sticks around the tent camp and Ertsberg. He didn't know it until later, but these were Amungme taboo sticks.

"I didn't know what this was about, so we unloaded our stuff and started putting tents together," Darnell said. "I had a shotgun under a blanket, but that was the only weapon we had. This was supposed to be a peaceful expedition, but what were these sharp sticks? This was moving me. They had poison vines hanging on them, snake skins and stuff like that. I took that it wasn't an invitation."

91

The men went ahead and set up the tents and returned to Timuka, where they were stuck for a few days because of rain. When they returned on the next clear day, they found more of the sticks, sometimes called *salib*—Indonesian for the Christian cross. The men continued the work of setting up the camp, but Darnell was somewhat unnerved. About a month later three or four Irianese men showed up at the camp, just staring at the miners.

"We looked at them for a while and nothing happened, so I walked up to one of them and held out my hand," Darnell said. "He didn't shake hands, but took the last joint of my middle finger between two of his knuckles, and pulled back hard, making a popping sound. So I learned the local handshake right then and there. They talked to me and I talked to them, but nobody knew what the other said. They were shaking from the cold and they were naked, so we gave each one a blanket and opened up a few cans of stew and fed them. Finally I told them they'd better go and they got the hint and left. They took some of the sticks with them, so this was a pretty good sign we weren't going to be molested."

Darnell found out from some missionaries what the sticks were, and radioed Darwin and asked Wilson what to do. Wilson's advice was to find Moses Kilangin, who had been so helpful on the earlier expedition. In 1967 Kilangin was living in Akimuga, where he was the head of the village, and Darnell and deputy superintendent John Currie, who could speak Indonesian, took a helicopter to Akimuga to get Kilangin's help.

Although it took some serious convincing to get Kilangin in the helicopter, they finally persuaded him and the three flew up to the Carstenszweide base camp to look at the sticks. Kilangin said that yes, they were taboo sticks, and they would have to go down and talk to the people of the Wa Valley, who were the most likely source of this gesture. So they flew down the valley and landed on the gravel bar in the Aghawagong in front of Nargi hamlet.

"Don stayed in the helicopter—I told him to keep the rotor going on idle," Darnell said. "We didn't know what we were going to have. The shotgun was under a burlap sack on the floor. The village was deserted, but pretty soon a kid stepped out of the jungle—I guess about eight or ten years old. He stood there and I looked at him and he walked right up to me and shook hands. He looked at me, and looked at the helicopter. About that time three men stepped out with

FREEPORT ARCHIVES, 1972

At the time it was built, the tramway from the mine to the mill was the longest of its kind ever built. It carried the first Ertsberg ore in 10-ton loads to the mill.

spears and bows, and John stood up, and Moses stood up and started talking. He talked to these guys for a few minutes and they laid their bows and arrows down at the edge of the jungle.

"In the meantime two women came out of the jungle and the kid called them and they came over where I was. And I turned around to the kid, who pointed at the helicopter. He wanted to get in the helicopter. So I boosted him up into the front seat. He sat in there jabbering away at Don, and Don didn't know if he liked this or not.

"The women came over and sat down in front of the helicopter, looking up at their images reflected in the Plexiglas cover on the nose. I got the kid back down out of the helicopter and he had a big sore on his leg, so I broke out the first-aid kit. I smeared him with sulfa and cleaned him up and put a Band-Aid on him. One of the women came up and pulled the Band-Aid off and stuck it in her mouth and was chewing it. I got her to understand that she shouldn't do that and got it back. Then the women started showing me injuries all over them. They've got a grass skirt about six inches long, you know, and nothing under it and nothing over anything else."

Darnell put Band-Aids on the women, and gave each a cigarette, which they chewed, rather than smoked. In the meantime the village chief—Tuarek Narkime, still a chief of Banti today—Kilangin and Currie were engaged in a heated discussion. The grievance, it turned out, was that the villagers were upset with the disruption of an area they considered special. Also, following the 1960 expedition, six villagers died, and Tuarek felt that the presence of the outsiders had something to do with it.

"We didn't know anything about this," Darnell said, "but I realized it was time to do something in the way of gifts." They had brought a sack of salt, tin cans, screw-cap jars, and steel *parangs* [machetes] and axes from the coast. Tuarek got the prize: a little Plumb half hatchet, with a hammer on one side and a hatchet blade on the other. "Boy, that did it. He was in seventh heaven with that thing. So we wound up the meeting and, through John, promised that we'd bring some more salt and other goods if they cleared a landing area for the helicopter. So we blasted off and that was the end of that meeting. We were safe. We were going to build our camp and then drill. That's what we were there for."

So they built the camp. The biggest problem was that Jacques Dozy's "alpine meadow" was, of course, a high-altitude swamp. As Darnell put it, "Wilson could've done me a lot of good if he told me it was a swamp. He forgot. I found out." Coping with the lack of solid ground cost the crew two weeks. Even though Darnell located the camp as far up on Ertsberg's talus slope as possible, it was difficult to build on the wet, tussocky ground, and once the floor sections of the buildings hit the ground they sank. So a foundation of two-by-twelves was fastened to them in the lowlands, and they were flown up to the Carstenszweide in this form. When the housing for the drillers was complete, it was time to bring in the drills.

The prefab buildings were brought up in sections in a sling load, but the drill rigs were too expensive to risk losing in the forest. These were three Canadian-made Longyear 38 units, each powered by a Ford Industrial 6 diesel, and they cost 38,000 good, hard 1967 dollars a piece. The rigs arrived by LCT on December 1, and were disassembled at Timuka into pieces small enough to make the trip to Ertsberg in the cab of the helicopter. While the Longyears were on their way to the highlands, two dozen Canadian diamond drillers were brought in to run the rigs.

The rigs were reassembled on skids in the *weide*, and dragged into position on the Ertsberg. Wilson and Del Flint had come up as the camp was being built. Wilson—then almost 60 years old and a Freeport vice president—made quite an impression on the Australian construction crew by working along with everyone else to get the prefabricated buildings, called "dongas," erected. Wilson and Flint, who ran the drill camp while Darnell looked after the base camp on the coast, picked out positions for the drill holes, essentially a 100-meter spacing along the base of Ertsberg. The plan was to drill right through the mountain, intercepting the ore body, and to continue for 20 meters into the limestone in the background to be sure everything was sampled. Five stations were set up, with the same sequence of holes, to yield a complete picture of the extent of the structure. About 200 holes were planned.

At first the drillers, who were briefed by the New York office, had expected to be working on very hard rock, and with new rigs that were difficult to run.

"I never will forget the first driller, French Canadian out of Quebec," Darnell said. "He couldn't talk English and I couldn't talk French. So I talked Spanish to him most of the time. We got by. Anyway, he set the drill up, and I stood right there. I had a Brunton compass, and I set the bearing and the angle. I set the first hole straight through. He started the drill, and I sat back on a rock and I looked over and that damn spindle, it just went in smooth and easy. I looked over at the driller and he rubbed his hands, saying the Spanish version of 'This is going to be a piece of cake.'"

On his first shift on that first hole, the driller registered 60 meters. New York had estimated 14 meters a shift. The drillers, who were paid by the meter, not by the hour, and with a bonus on top of that, were thrilled. And this caused some trouble at first. With such soft rock, and new, powerful rigs, the men just shifted their machines into top gear and went for broke. But as Darnell put it, "Their idea was to make a lot of holes, but we were there for core." And running a drill at full speed is not the way to get useable core.

"I told 'em, 'You get anything less than ninety-two percent, you're going to re-drill the hole on your own account,'" Darnell said. "This registered, and we had no more trouble after that. Instead of drilling in high gear they were drilling in second. This slowed things down, but we were getting the core, which was the whole idea of the thing."

The Indonesian government wanted the company to bring in four Indonesian engineers to train in field identification of core and core logging. Two spoke English. One had a masters degree and the others were either engineers or geologists. "They were all crack people," Darnell said, "they just needed field experience."

At the time, the core shack was just being established. The jaw-crusher, the pulverizer, the screen system and the splitting system were all in place. The process requires reducing three meters of core to a quarter-pound sample of ground rock. In the case of the Ertsberg core, Freeport had arranged to take just half the core, leaving the rest with the government to file, so all the core had to be split lengthwise. This way if the company decided not to proceed with developing the deposit, the government would have an independent record of what Ertsberg contained, a system that is still used today.

Darnell showed the engineers how to run the crushing and pulverizing equipment, how to screen down the crushed core, and how to label and log the samples. The young men were smart and easy to work with. But as it turned out, they were from pretty high-class families, which threatened to cause difficulties.

"I ran into this problem," Darnell said, "when I started telling them, 'Look, in order to show somebody else how to do something, you have to know how to do it yourself. I'm going to show you how to do this, and you're going to run the machine. When I think you know how to do it, you teach this Irianese man here how to do it, then *you* supervise.' And they objected. They said, 'Look, we're college graduates, we don't do manual labor.' Well, it took me a few minutes to get organized. That night I called up Darwin and asked for gloves. They came in on the next boat and I put white gloves on them. Engineers with white gloves. I told them, 'That's a *status* symbol, nobody else wears gloves, not even me.' That did it. From then on there was no trouble. They were good men all of them. And they learned a lot, they learned how you do it. That was the key."

Once the core shed was operational, and core was moving smoothly through, Darnell started on another project he had been given by Wilson: scouting possible sites for a port, an airport, a town, a mill, and an access road, and looking for sources of hydropower. He never had time to scout a location for the mill, and saw no obvious sources of hydropower that would continue through the dry season. But Darnell was the first to suggest a port site on the lower Aikwa

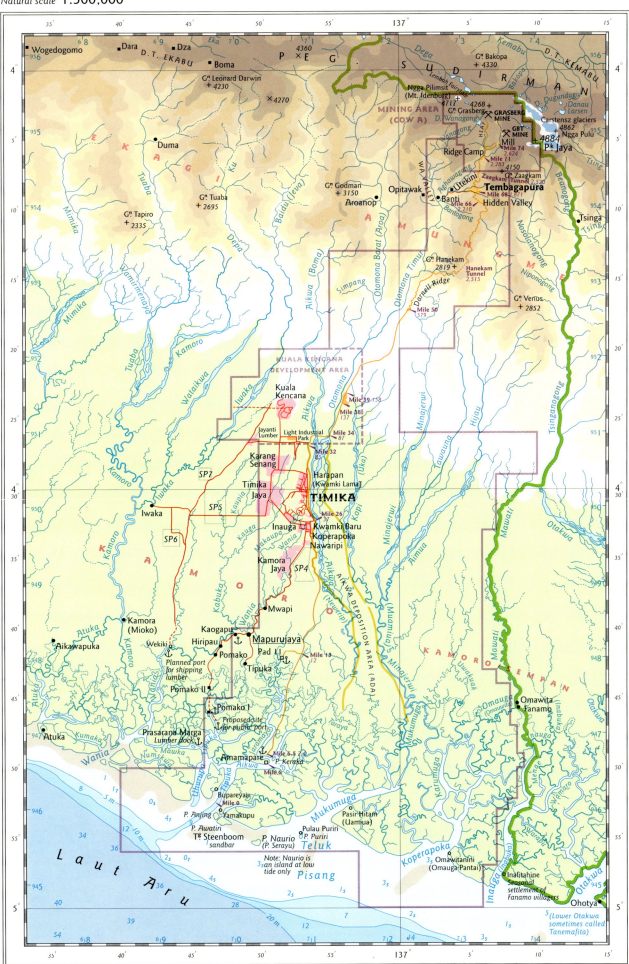

The foundations for the tramway from Ertsberg to the mill are nearly complete in this 1972 photograph. Note Grasberg looming in the background.

(later Amamapare), an airport on the level land near the middle Aikwa (Timika Airport), a town in the valley of the Mulki and Uteki Rivers (Tembagapura), and a workable route for an access road.

Darnell first took an aluminum launch with a crew of Australian fishermen and surveyed the rivers in the area. The Tipuka–lower Aikwa showed the most promise, as soundings showed 8 meters over the sandbar at the mouth, and 12 meters in a channel continuing for a good 6–7 kilometers upriver. Although it doesn't sound like much, this is the deepest running river in the area. Then he went up in the helicopter scouting for airport sites. They already had their safety pad, which was a gravel bar upstream from where the airport is today. Relying on advice from La Freniere, who was once a major in the U.S. Air Force and who piloted John F. Kennedy on his presidential campaign, Darnell took photographs of several possible sites on dry, level ground in and around the middle run of the Aikwa.

Then he started looking for possible town sites. Darnell had extensive experience living and working in the tropics, at 2,500 meters, at 1,200 meters, and at sea level. "In the tropics, wherever you are, the climate depends on how high you are," he said. "You're on the coast, it's hot; you go up high, it's cold. Somewhere in between there's a happy medium. I found that six thousand feet [1,800 meters] anywhere in the vicinity of the equator, usually you find springtime. So we started looking for this with the helicopter." He flew down from the Ertsberg, over the Aghawagong, to the settlement at Nargi. Then he saw the valley just east of there, and dropped down to tree-top level to get a reading on the altitude. It was surrounded by rock cliffs, and at just about the ideal elevation, so he marked it on his map: "possible town site." Today this is Tembagapura.

"The last thing on the list was the road," Darnell said. "And we looked, and we looked. We even went behind what's now Tembagapura and tried to find a way back up in around the mountains west of the Ertsberg. We never could get in. We got into some pretty tight canyons in there, but never found anything that could go with a road."

One day, frustrated as usual, Darnell and La Freniere were resting down at their safety pad near the airport site. It was common for them to stop there while Darnell wrote up his notes and recorded the subjects of his many rolls of film. Sitting on the gravel bar, the men had an inspiration.

"Don said, 'Look at that ridge over there,'" Darnell recalls. "And I thought, Well, so what? It's a ridge. And then the light came on in my head. You can't built a road on a forty-five degree angle slope. But if you take the damn top off that ridge and shove the dirt both ways from it, you can make a grade up there."

So Darnell had Flint and La Freniere photograph the ridge. With the door of the cab removed and Flint tied to the skid, La Freniere flew at tree-top level while Flint took 350 photographs of the ridge. They ran into the wall at Hanekam, where the tunnel is now, and picked up the ridge on the other side. From there, Darnell reasoned, a roadway could be build down to the proposed town site. He wrote up a report, added his pictures, and sent the package on to Wilson. Later, this route would be the same one used by Bechtel and J. H. Pomeroy when they built the access road, and even today the ridge is called "Darnell Ridge."

BY THE TIME the drilling program was finished, more than 7,500 meters of core had been taken out. Some of the holes had probed 360 meters into Ertsberg. Unfortunately, assays showed that the original estimates based on Wilson's 300 kilograms of surface samples—3.5 percent copper and perhaps 50 million metric tons of ore—were overly optimistic. After drilling, estimates dropped to 33 million tons of ore, at an average grade of 2.5 percent copper.

This was bad news. Most mines would be very happy with an ore body of 2.5 percent copper, and many profitably work ores of less than 1 percent copper. But most of these mines, as Wilson somewhat laconically states, "are more conveniently located." Lying in such rugged terrain, and so far from existing infrastructure, the smaller than expected ore body and the lower copper grade almost killed the Ertsberg mining project.

But Wilson was not about to abandon his pride and joy. The next phase, a feasibility study, would cost at least $2 million. Several of Freeport's directors and officers thought, quite rationally, that this would be a waste of money. Theirs was a conservative company which had almost never borrowed money. Ertsberg would need huge amounts of cash to develop, and the size of the deposit might not justify the expenditure. Some at Freeport thought Wilson was being way too optimistic.

Wilson felt that the geology of the area ensured that Ertsberg was not an isolated deposit, but he needed proof if he was to get backing from the board. He sent Del Flint back to Irian to look more carefully for additional prospects. Flint first examined a limestone cliff near Ertsberg, where Wilson had noticed stains of green malachite in 1960. There he found obvious, strong surface mineralization in an area measuring about 1,000 meters by 300 meters, very close to Ertsberg. He also noted an outcrop of ore to the southeast at over 4,200 meters elevation, just under a dramatic limestone mountain Jacques Dozy had named Dom, or "cathedral" in Dutch.

Armed with Flint's field report, Wilson drew up a proposal for the further development of Ertsberg. In October, the board agreed to hear his proposal. The general consensus at the beginning of the board meeting was negative, and Wilson began to think seriously of resigning. But then Augustus C. Long, a board member who had for many years had been the chief executive of Texaco, spoke up. Long noted that the sulfur market, at the time 85 percent of Freeport's business, was weak. He pointed out that other than Ertsberg, the company had no other major projects on the drawing board. There were risks, but Wilson's Ertsberg promised good rewards. Long swayed the board, and Wilson got his $2 million for the study.

Wilson immediately flew to San Francisco to meet with representatives of Bechtel and J.H. Pomeroy, the two companies he had chosen to do the work. Bechtel is a huge engineering and construction firm that even by the late 1960s had successfully carried out a series of tough projects throughout the world, and Pomeroy was a smaller operation specializing in marine installations. Despite their experience, the advance party, once they saw the terrain, reacted with awe and amazement. Later, Bechtel executives would say that the Freeport project was the single most difficult engineering project they had ever undertaken.

With traffic getting heavier at the project area, the company could no longer rely on the long and difficult boat ride from Darwin. Wilson himself almost went down in a cyclone once on the unpredictable Arafura Sea. Without an airport, Freeport needed a seaplane, and one with a good range and substantial capacity. Unfortunately, no commercial seaplanes meeting this specification were being manufactured by the late 1960s, and the only option available was to find an old World War II–era U.S. Navy PBY. The problem with these used war survivors was that none could be found in still serviceable condition. By sheer coincidence Wilson met a man who owned a PBY that

Amungme living near Tembagapura, awaiting the arrival of President Suharto, celebrate with Indonesian flags. On his visit he renamed the province Irian Jaya.

sat unused in a hanger in Oakland, California. Freeport's chief pilot, Dick McNally, happened to be in San Francisco at the time and hustled across the Bay to take a look.

This plane was in perfect shape, probably the only one in the world in such condition. It had belonged to the eccentric billionaire Howard Hughes, who had intended to use it to go salmon fishing in Alaska. Once the old engines were replaced and the luxury accommodations and wing-fitted fishing dinghies were removed, the plane was ready to begin its new life ferrying Freeport personnel between Darwin and Timuka.

Wilson's next problem was to bring out enough sample ore from Ertsberg for a large-scale metallurgical test, to determine if the composition of the ore would allow it to be economically processed, and if so, to determine the proper mill design to process it. This kind of test ideally requires a sample of 10,000 tons of ore. In Irian, however, this was out of the question, as it would have required 20,000 helicopter trips, which Wilson calculated would take perhaps ten years. Although it was a bit riskier, they decided 300 tons would do.

Even 300 tons wasn't easy. In the interest of time, the company hired one Australian miner, and a crew of ten young shrimp fishermen from Darwin. Although, as Wilson writes, the fishermen had "never seen a pneumatic drill or a stick of dynamite," they had strong backs and were willing to work. The men soon drove a 100-meter adit into Ertsberg. They extracted the required ore, and tests demonstrated that it could be cost-effectively milled and concentrated.

Bechtel and Pomeroy provided their first feasibility report in August. Their plan was based on mining Ertsberg by open pit, crushing and grinding the ore near the mine, mixing this with water and piping the resulting slurry to the coast in a large pipeline, and concentrating the ground ore at a facility on the coast. This method would yield two concentrates: one of copper, with associated traces of silver and gold, and one of iron. The total cost of this facility would be $160 million. This was too much money for the company to justify, and since pig iron sold at the time for just $200 a ton, it was decided that recovering the iron would not be commercially viable. Bechtel-Pomeroy revised their proposal, locating a combined crushing plant and concentrator near the mine, and using a much smaller pipeline to bring the concentrated copper to the coast. This would be considerably cheaper, costing an estimated $100–$120 million.

Then came the toughest part: finding investors. There was almost nothing routine or average about the Ertsberg project, and although this excited mining engineers, it did not have the same effect on investors. For most of 1969, Freeport Minerals vice-president Paul Douglas worked to secure long-term sales commitments from copper smelters in Germany and Japan. Company treasurer Nils Kindwall, an early and enthusiastic supporter of the Ertsberg project, flew worldwide seeking loans. Through the efforts of these two, an intricate $100 million loan package was assembled with the participation of thirty different lenders.

Approximately $20 million of this package came from a Japanese concentrate trader and seven Japanese smelters, who together agreed to buy 65 percent of the production over the next 15 years. Another $22 million came from Kreditanstalt für Wiederaufbau, the German development bank, with a contract from Nord-deutsche Affinerie in Hamburg, Europe's largest copper smelter, to purchase the rest of the ore concentrate for the life of the mine. An additional $40 million came from a consortium of seven American banks, with their investment guaranteed by the federal Export-Import Bank. Freeport itself anted up $20 million. Provisions for cost overruns were built in, which proved wise: by the time the project was finished, because of inflation, engineering modifications, and the incredible difficulties encountered building the access road, the original $120 million had swelled to $200 million.

The 101-kilometer access road, from the swamps of the Jaramaya River to the mill site in the valley below Ertsberg, was the most difficult aspect of the construction, and in fact is almost assuredly the toughest road-building project attempted anywhere in the world. Wilson, the feasibility engineers from Bechtel-Pomeroy, and anyone else who had been to the area knew the road would be tough, but nobody was prepared for the almost limitless ability of the New Guinea landscape to hinder the road-building effort. The land also had an expensive appetite for equipment. One of the tractors was buried in a landslide. A bulldozer toppled off into a 700-meter ravine. Two trucks fell into rivers. One helicopter blew off a cliffside helipad; another sank into a swamp.

In building the section of road from the landing pad on the Jaramaya River (Pad 11) to about where Timika is today, the contractors dumped endless truckloads of river gravel, trying desperately to create a sub-grade in the swampy muck. At its worst, this section sometimes required 50 cubic yards of gravel to advance the road just one foot. The engineer in charge of this construction was at first puzzled that the sum of his daily logs of completed footage did not match the total length of the road. Then it dawned on him. At night, after the men knocked off work, the leading edge of the road sunk back down into the swamp. In the worst section, the crew actually laid down two miles of road for every one mile that now stands.

The easiest section, it was thought, would be the level area between Mile 39 and Mile 50, above the tidal swamps and yet below the mountains. A man could usually walk on this ground, but beneath the superficially solid surface was a three-meter layer of peat, dead vegetation and clay with, Wilson notes, "the consistency of grease." This material could swallow bulldozers. The crew took to taking soundings using a long pole, laying their road where the "grease" was thinnest.

From milepost 50 onward, the road followed the "Darnell Ridge" up into the highlands. As Darnell surmised, following the ridge top

President Suharto, at center, seated next to Freeport Minerals president Forbes Wilson at the dedication of the Ertsberg mine in March 1973.

was the only way to get the road inland. Valleys were prone to flooding, and generally ended in headwalls. The slopes were far too unstable. These are young mountains, and the process of erosion is still very active. In many areas the crest was knife-edge sharp. Tiny Caterpillar D-4 bulldozers, barely larger than riding mowers, were first choppered in to shave off the edge. These were followed by larger D-6s, and then D-7s and D-8s. Finally the 25-ton D-9, with its 7-meter-wide "heavy iron," was brought in to finish the job. Throughout the construction, the grade was maintained at no more than 25 percent, which in places has resulted in elaborate series of switchbacks.

Two mountains resisted the engineers' efforts to find a way around them: Mount Hanekam, about six kilometers past Mile 50, and Mount Zaagkam, between Tembagapura and the mill. The 2,819-meter-high Hanekam has such a distinctive shape that it was independently given the same name twice. The Dutch Otakwa expedition under J.J. van der Bie named it "Hanekam," which means "cockscomb" in Dutch, and the British Ornithologists' Union Expedition under Walter Goodfellow named it "Cock's Comb." This obstacle

Freeport consultant and later P.T. Freeport Indonesia president Ali Budiarjo, at left, with President Suharto. Budiarjo was crucial to negotiating the contract of work.

required Bechtel-Pomeroy to bring in 100 Korean coal miners to drive a 1,100-meter tunnel. An 845-meter tunnel also had to be cut through 4,150-meter Mt. Zaagkam, which looms over the valley in which Tembagapura nestles.

The almost vertical headwall between Ertsberg and the valley of the Aghawagong, which gave Dozy so much trouble in 1936, proved an effective block to further road-building. The mill would have to be built at its base. To get to the ore a two-unit, single-span tramway was built, rising 680 meters over a horizontal distance of just 1.6 kilometers. Once the structures at both ends were completed, helicopters flew one end of a 3,000-meter polypropylene rope to the top, and back down again. To this, progressively heavier cables were affixed, until the final working cables, as big around as a man's arm, could be winched up. At the time of its construction, this was the longest aerial tramway in the world, and a major problem developed that had not been seen on shorter trams. The 10-ton ore buckets made the cables oscillate so violently that the buckets were sometimes derailed or flipped off. Our engineers tried reducing the load, slowing down the cables, and lowering the center of gravity of the buckets, all to no avail. Finally a Swiss mathematician was located who had studied tramway oscillations. He performed some complex calculations, then ordered small modifications that brought the oscillations under control. "A tramway is like a violin," he said. "It has to be tuned."

When it was first constructed, the 109-kilometer slurry pipeline, which follows the road from the mill to the coast, was also the longest ever built. The concentrated copper slurry in this line—there are now several—must be pumped, as it is not a continuous downhill drop from 2,800 meters at the mill to sea level at the coast. Copper slurry is extremely abrasive. If pumped too fast, it eats right through the steel pipeline; too slow and it clogs up, requiring the pipe to be replaced. Elaborate computer calculations and lab testing came up with the optimum combination: a slurry mix of 64–67 percent water, and a pumping rate of just over 5 kilometers an hour.

BY THE END of 1972, the road was finished, the trams were working, the mill was operational, and the slurry line was in place. In December, the first 10,000 tons of Ertsberg concentrate headed for Japan. The mine was up and running. Three months later, President Suharto arrived. He rode by jeep up the magnificent road, and the next day named Tembagapura—"Copper City"—and dedicated the new mine. He also, unexpectedly and on the spot, renamed the province Irian Jaya, or "Victorious Irian."

PART II
Crisis and Expansion

An ore bucket returning to Ertsberg for another 17 metric tons of ore in 1983. Tram capacity eventually proved a bottleneck, and ore is no longer transported this way.

CHAPTER FOUR

The Underground Era

Marginal profitability—Steepening the pit—Vexatious deposits—Ertsberg East—The Filipino miners—A stubborn block cave—The ore passes—Gold fever in the boardroom—A spectacular core sample

WHEN I FIRST joined Freeport in 1975 as manager of mine development, the company was facing a major turning point with the future of its Indonesia operations. At that time, Ertsberg had approximately 12 million tons of remaining open pit ore reserves, which the mill was consuming at an annual rate of 2.7 million tons. The operation was, at best, marginally profitable. The board of directors had grave concerns as to whether or not it would be able to pay back $105 million in outstanding loans from international banks. This was a staggering debt for a company that was operating with $74 million in gross revenues and $6.8 million in net income. Remaining open pit life was a little over four years. It was hoped that the life of the mine could be extended by going beneath the pit, using underground mining methods to get to the roots of the Ertsberg deposit. My first task for Freeport was to determine whether or not this could be done.

In late September of 1975, I went to Indonesia for the first time, a visit that will be etched in my memory forever. I knew something of the mine and of the town of Tembagapura. I knew the logistics were almost unbelievable, involving barges traveling up twisted, brackish rivers, a 63-mile road through swamps and nearly impossible mountainous terrain, and finally an aerial tramway shooting up a 680-meter, near-vertical rock face to the mine. I knew that the construction of a mine, a concentrating plant, and town of 1,200 people in the mountainous interior of Irian Jaya was a miracle. But one does not truly appreciate this striking combination of engineering accomplishment and raw beauty until one sees it firsthand. I was dumbstruck. These were the most rugged mountains I had ever seen—and I was raised in Alaska—and there was copper mineralization everywhere. I knew I had made the right decision to join this company.

It did not take long to determine that going underground to get at the remainder of the Ertsberg deposit would not be economically viable. The combination of low metal prices, high capital costs to develop the mine and train the workers, and the high cost of extraction made the undertaking impossible. The mine operators did have an alternate plan, however, and I thought it showed promise. This was to reach the remaining ore by making the pit deeper. There were drawbacks to this proposal. To make the pit deeper we would have to steepen the pit walls, which would require considerable, and costly, stripping of waste before the company would reap any benefit from the newly exposed ore. Furthermore, steepening the wall angle—we planned an increase from 45 degrees to 55 degrees—created a potential for the collapse of the pit walls.

Freeport hired Pincock, Allen, & Holt, a mining consulting firm based in Tucson, Arizona, to study the feasibility of this approach. After several trips to the mine over a six-month period, the company's senior rock mechanic engineer, Rick Call, and his staff concluded that, indeed, the pit walls could be safely and economically steepened. This strategy would allow us to mine an additional 6 million tons of the 10 million tons of ore we knew were lying below the current pit limits.

In late 1976, armed with this information and Call's technical assistance, Freeport Indonesia Executive Vice President Milt Ward and myself presented our proposal to the board. Milt was the person most responsible for convincing me to leave a good job with Fluor in California, and move to Louisiana to work for Freeport. He was, and still is, a very persuasive person. That and my knowledge of his very considerable skills as a manager and engineer—he had previously worked for Ranchers Development and Kerr-McGee—helped convince me to make the move.

The board meeting at which we proposed our pit-steepening plan was rather fiery. Members groaned at the amount of up-front cash required. The lenders, who at the time sent representatives to the board meetings, protested the financial risk involved, and the length of time it would take to repay the loans required to finance the initial stripping. One bank brought its own mining engineer to the meeting. He loudly proclaimed that it was sheer folly to steepen the pit, and that the walls would most certainly fail. His analysis was based on a review by Bechtel, which unfortunately for us was a very respected firm that had built much of our original infrastructure and production facilities, and had produced the original feasibility study.

One director, who had ties to our customers in Japan, even suggested that we shut down the entire operation for a year or so, until copper prices recovered. There was certainly some logic to this, as a glut of concentrate and copper metal on the Japanese and world markets had seriously depressed copper prices. However, demobilizing and remobilizing 700–800 workers and their families, then restarting a plant left sitting idle for a year or two in Irian Jaya's tropical climate was regarded by Freeport's operations managers as, to put it mildly, rather impractical.

Rick Call remained calm throughout the meeting. He has extensive firsthand experience with mines around the world, and at the time was a professor of rock machines at the University of Arizona. He did a masterful job of defending the technical side of the plan and in the end the Freeport Indonesia board, with the concurrence of the banks, adopted the plan. This decision, although it involved just 6 million tons of ore, was one of the most important in the history of the company. It gave Freeport the time to look for other ore reserves, and extended the life of the project. If we had not been able to add open pit life, I am certain the mine would have been closed—before we discovered the rich deposits east of Ertsberg, and before Grasberg. The $8 billion company we see today would simply not exist.

But the battle was not over yet. When the Indonesian ministry of mines learned of our plan, they objected. I found myself racing to Jakarta to explain our actions. As the body charged with managing Indonesia's mineral resources, they argued that by embarking on a program that would recover just 60 percent of the remaining ore, we were wasting 40 percent of their asset. We countered by arguing that it would be better to mine 60 percent of the ore by open pit than none by underground. Even if cost were no object, we pointed out, an underground mine could extract no more than 70–85 percent of the remaining Ertsberg ore.

Finally we were able to convince the ministry. In the end it proved to be a very good decision from their point of view. While proceeding with the pit, we discovered an additional 2 million tons of recoverable ore in the fringe areas of the deposit that would never have been found if we had gone underground. And, of course, Indonesia now

KAL MULLER, 1985

An Irianese miner working in the Ertsberg East—or Gunung Bijih Timur—underground mine. This was a difficult mine to bring into production.

has an ongoing mining enterprise in a mineral district that will still be in operation 100 years from today.

Once the pit extension plan was in place, the opportunities for expanding the ore reserves were excellent. The geological setting of the Ertsberg district is such that one finds copper mineralization in every nook and cranny. Freeport's original contract of work, universally called the COW, is a 100-square-kilometer block centered on Ertsberg that lies over an area marked by extensive faulting.

The Ertsberg deposit is a skarn ore body, or a "contact metasomatic deposit." That is, it is an ore body formed not within an igneous intrusion itself, but rather one that formed when mineralizing fluids released by the intrusion contacted, and subsequently altered, nearby deposits of limestone. The process began 3.25 million years ago when magma forced its way up from the earth's mantle under the 36–53 million-year-old limestone that makes up the mountains of the Ertsberg district. This and subsequent similar intrusions, characterized by very high temperatures and pressures, produced extremely hot and active fluids, rich in copper and gold, that forced their way up through cracks in the limestone. In the places where they contacted the limestone, these very high-energy fluids altered its chemical structure, leaving behind a skarn, characterized by copper- and gold-containing minerals.

In 1975 I was not familiar with mining skarn deposits, and until then had worked mostly in porphyry copper mines. Porphyry is a more common form of copper deposit in which fractional crystallization of an intrusive magma, and continued or repeated intrusive activity, leads to a concentration of copper-bearing minerals within the area of the intrusion. With porphyry copper, one mines the original intrusion; with skarns, one mines scattered, enriched "pods" of skarn surrounding the intrusion. Finding and mining these skarn bodies can be a challenge, as my college textbook states:

> The deposits generally consist of several disconnected bodies. They are mostly small as compared with "porphyry coppers," or sedimentary deposits. They are vexatious deposits to exploit because of their relatively small size, their capricious distribution within the contact aureole, and their abrupt terminations. Like the scattered plums in a plum pudding, they are difficult to find; they exhibit few 'signboards' pointing to their presence, and costly exploration and development are necessary to discover and outline them. Their development must be undertaken with caution, and the optimism attendant upon mining such concentrated and often rich bodies frequently gives way quickly to disappointment upon the sudden termination of the ore body.*

The textbook suggests that operations would not be easy with these types of ore bodies, and indeed they weren't. The Ertsberg itself was unusual for a skarn deposit, because it was very large and even a little predictable. Conveniently, for the first few years the ore we were mining projected above the ground, so we knew very well what we would be mining. When the operation became a pit, it ceased being a simple matter to determine the extent and shape of the ore body. The pit did, however, give us the opportunity to drill for small satellite ore bodies that were unreachable from the surface.

OUR GEOLOGISTS KNEW from the beginning that Ertsberg couldn't be the only ore body in the area. As early as 1968, while diamond drilling was still going on to determine the extent and richness of the Ertsberg ore body, Del Flint had discovered an outcrop of

---

*Alan Bateman. *Economic Mineral Deposits* 2nd ed. John Wiley & Sons, Inc., 1954.

ore just east of Ertsberg. Flint, Freeport's chief field geologist at the time, had been directed by Forbes Wilson to conduct some preliminary exploration to convince the board that they were dealing with a mineral district, not just a single deposit. In 1973, once the mine was producing, geologist Frank Nelson was brought in to map and explore the district. He began with the ridge just east of Ertsberg, which he called the Flint Extension.

The Flint Extension was about a kilometer and a half east of Ertsberg. Although this sounds close, in those days, with no roads, getting there required a tough hike through swamps and up a steep grade. Frank spent the first two years taking surface samples and mapping the district just east of Ertsberg, and then began drilling his targets in 1975. He was the perfect geologist for this type of work. He had enthusiasm and an extensive knowledge of geology.

When I first visited the site in September, 1975, Frank was just completing the drilling of a very high-grade surface outcropping called the West Hanging Valley, just a few hundred meters southeast of Ertsberg. He had found some small, high-grade pods of skarn almost directly above the mill. From the site you could look right down to the base of the tramway.

At the time, Freeport was looking especially for reserves that could be mined by open pit, as the general belief within the corporation was that the technical and economic difficulties associated with underground mining at this location would render an underground ore body economically unviable. The West Hanging Valley, with its rich surface outcropping, looked like a good candidate. Unfortunately, although the grade approached a remarkable 15 percent copper, drilling revealed that the deposit contained just 75,000 tons of potential ore, and, to make matters worse, the material would have to be mined from underground.

Frank started concentrating on the area further east, where fresh sulfides could be seen right on the surface. His drill platforms perched precariously on the steep terrain, and water for the rigs had to be pumped up through pipelines from streams sometimes 300 meters below. The whole operation was supported by helicopter, and the weather, altitude, and terrain severely taxed the capability of these aircraft. The drill towers of the Longyear 38s were assembled and disassembled while being supported by a hovering helicopter, a very delicate procedure. Luckily, there were no accidents.

Frank hired local highland Irianese to support his crews, and they proved ideal workers under conditions that, though harsh to outsiders, were familiar to them. When they encountered rough, difficult terrain, they would take off their boots and clamor over the obstacles in their bare feet. Eventually, Frank's drive and enthusiasm paid off, as he later recalled:

> With those drills we were able to just touch what we found to be a very good ore body. At the very deepest parts, we were getting down into a mass of bornite and primary chalcocite. But that was as far as the machines would go, probably five hundred meters. The limiting factor was pulling the rods, not turning, but of course the driller had to have superchargers because of the altitude. It was so tantalizing; we were ending in beautiful-looking ore. We got so spoiled that if the core just had chalcopyrite in it, you'd say, "Hell, its just chalcopyrite." A pure chunk of chalcopyrite would be about thirty percent copper; a pure piece of bornite would be about sixty percent copper and a pure piece of primary chalcocite would be in the range of eighty percent. You got to where the chalcopyrite was almost regarded as bad luck.

The ore body Frank Nelson intercepted in 1975 became known as Ertsberg East, or Gunung Bijih Timur—"East Ore Mountain" in Indonesian—now referred to simply as GBT. This deposit later was shown to have two additional zones: the Deep Ore Zone, or DOZ, and the Intermediate Ore Zone, or IOZ. Bornite is the main copper mineral at the GBT mines, whereas the original Ertsberg was mostly chalcopyrite. The Ertsberg East complex lies along an ore-controlling east–west structure, the Hanging Wall Fault Zone. This structure forms the boundary between the ore-bearing skarns to the south and the barren marble and siliceous hornfels to the north. It can fairly be said that Frank's discovery of the GBT ore body assured the continuation of the company.

Money was tight during this period, and Freeport found it difficult to support this drilling program in the manner it really deserved. Frank, however, was an expert scrounger. He supplied his program using resources cast off by the miners and operators. Old shipping containers became offices. Pipe for the mine somehow found a home in his operations. His program was always well-stocked, including a regular supply of T-bone steaks for his Irianese workers.

The surface drilling continued through 1976, by which time most of us were certain that a significant ore body existed. However, it was an underground skarn ore body. It was too deep to be mined by open pit methods. This was not considered good news by Freeport's senior managers, as the company had never undertaken an underground mining operation before. They were, however, familiar with the risky reputation of underground mining.

Most of 1976 was spent analyzing this data as it came in. It appeared from the drill results that there were approximately 40 million tons of ore in this new deposit, averaging about 2.5 percent copper. Although these are good numbers, it was extremely difficult to demonstrate that this would be an economically viable deposit at a time when copper prices were stalled at 63 cents a pound. Also, the drill spacing was too wide to prove with certainty that our ore reserve estimate was correct. Because of the length of the holes, and the difficult access, it was obvious that a surface drilling program comprehensive enough to prove the deposit would be too expensive and time-consuming, so we investigated other ways that Ertsberg East could be both demonstrated and made commercial.

Our target for economic feasibility was a 20 percent return or better, and it wasn't until the eighth study, based on a mining method called "block caving," that the deposit began to look like it had commercial possibilities. Block caving is a low-cost method of underground mining in which large blocks of ore, say 120 meters by 120 meters in horizontal area, are undercut, thus allowing the ore to cave under its own weight. Once it has caved, the resulting boulders and rocks are broken up and sent to the mill for concentrating.

We proposed a two-step program to the Freeport Indonesia board. First, for a capital outlay of $12 million, we would drive a 1,300-meter adit from a point near the Ertsberg open pit into and through the center of our deposit, which we then called "Ertsberg East." From this adit we could cut a drill drift parallel to the mineralization, from which we could drill test holes on close enough centers to precisely measure the volume and grade of ore.

Once armed with this information, we could classify the material as ore reserves in accordance with U.S. Securities and Exchange Commission regulations, prepare bankable feasibility studies, and come up with an accurate cost estimate for the overall project. The second step, once we were able to get financing based on our drill data, would be to bring the mine into production, which we then estimated to cost $70–$80 million. We called this estimate an "order of magnitude" estimate because it was done with no detailed engineering, just our general knowledge of construction costs for similar work.

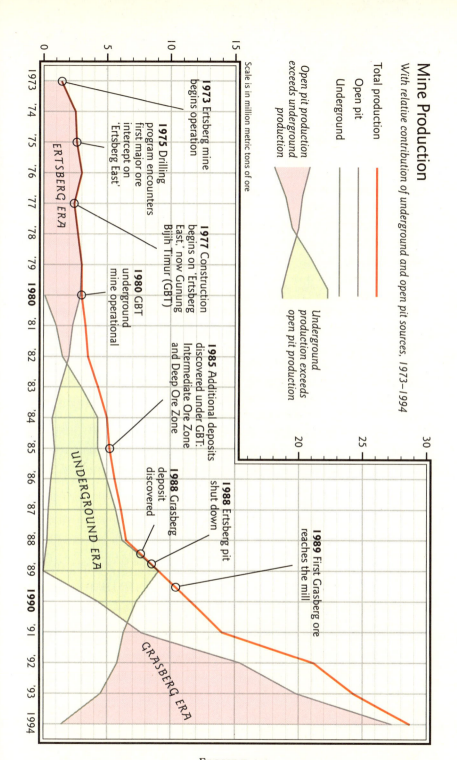

FIGURE 4.1

In December 1976, the Freeport board authorized the $12 million for the first stage of the project. This was a brave step for the company, which was hard-pressed for cash. However, as one board member commented at the time, "Either we do the project or go out of business in a few years." In January 1977, the first blast was taken out of the portal of the Ertsberg East adit. The contractor for this work was Cementation Engineering, a British firm specializing in underground construction and grouting work. Getting the crews and equipment to the site took just six weeks, and the adit reached the ore body in January 1978. By September, the portion of the deposit above 3,600 meters had been evaluated from ten underground diamond drilling stations.

Rather than wait for all of the drill and metallurgical data to come in, as soon as we had enough that the project looked viable, Freeport's mine development department—together with Bechtel—proceeded immediately with "bankable" feasibility studies, even while downward drilling continued. This kind of speed is typical of Freeport. Often the board would authorize some limited up-front capital to be spent for engineering and equipment orders—with cancellation clauses—which could shave a year or more off a project's completion time. In this business, time is money.

By November 1978, the feasibility study had been completed, including a plant-scale metallurgical test of the ore. Running samples through the mill showed conclusively that Ertsberg East ore produced excellent concentrate, and that when fed a combination of Ertsberg ore and Ertsberg East ore, the existing mill could be pushed from its design capacity of 7,500 tons a day to 9,500 tons a day. Better still, even though there would be many more tons of copper being produced than were originally contemplated, another concentrate pipeline to the coast would not be required. We calculated the overall cost of the project, including all contingencies, to be $101 million, leading to an operation that could produce copper at 55.6 cents a pound, for a 24 percent return on capital.

By May 1979, the company had contacted potential buyers of the additional concentrate, and lenders who would commit to loans for the project. Much to our pleasure, the lenders' mining engineers agreed with our assessment of the project's viability, and we therefore received considerable support from several international banks. By May 1979, we recommended to the Freeport Indonesia board that we proceed with the project—which would yield an excellent profit and would extend the life of the enterprise at least 10 years. Freeport, by then, had begun to take steps to hire skilled senior managers who were experienced in underground mining operations.

We first hired Les Acton, one of the leading authorities on block caving, and a master of large and complex mine management. Acton, who was hired after taking early retirement from his job as mine manager of Newmont's San Manuel mine in Arizona, was our assistant general manager on site during the drilling of GBT, and was instrumental in the final selection of block caving as the mining method. Once the project was approved, he became general manager. Gunung Bijih Timur would prove a very difficult ore body to mine, and Les is the one person most responsible for making it the success it became. Assisting him was a cadre of senior mine operators, including Joe Murray, who later became general manager, and Bill Bienemann. These men also came from the San Manuel operation, and both knew their business well.

Drilling the Ertsberg East deposit from an elevation of 3,600 meters upward had proved to be fruitful. By the time the decision was made to proceed with the project, we had determined that, from this elevation, we would have access by block cave mining to 36 million tons of extractable ore with an average grade of 2.3 percent copper. Better still, we saw excellent indications that the ore did not stop at 3,600 meters, but, in fact, continued to a considerable depth.

In September 1978, drilling continued at depth, and these additional deep holes demonstrated that the ore body extended down to the elevation of the mill, at 2,900 meters. For convenience, the mineralization encountered by the drills was divided into two zones, intermediate and deep. The IOZ and DOZ, combined with the substantial proved reserves higher up, gave backers of the expanded project considerable confidence that this could be a mine with a long life. By August 1978, loans had been arranged, and we began to mobilize the construction force to mine Ertsberg East.

The project gathered momentum in 1979, and the complexities of managing a difficult underground project of this type began to strain the corporate office in New York City. As a result, I was transferred from New Orleans to New York and promoted to vice president of Freeport Indonesia in charge of projects. My function would be to oversee the GBT expansion, and by midyear, I relocated to New York

City. The mine development department still reported to me through Dick Stewart. Dick, who had worked in Freeport's chemical, nickel and coal divisions, became the manager of mine development.

Probably the biggest difficulty in bringing underground mining technology to Indonesia was training the miners. In the United States and in other countries with a tradition of underground mining, there is a pool of people who are experienced in this type of work, and who have the skills to work safely around the potential hazards it presents. Indonesia does not have a tradition of mining, and there were no trained Indonesian underground miners to hire. It typically takes seven years to train an all-around miner, and until he has enough experience to work safely, he must be closely supervised, which makes such trainee workers expensive. On top of this, we found ourselves faced with training hundreds of Indonesians who not only had little desire to work in an underground mine, but were sometimes even afraid of it.

Our first approach was to hire a cadre of skilled hard-rock miners from the United States and Canada, hand-picked by our managers and foremen. This quickly turned out to be a disaster. The men were probably excellent miners, but they had character flaws that made them totally unsuited to our purposes. North American miners are a two-fisted, heavy drinking breed, and most are able to get themselves into trouble at the least opportunity. Of the forty hard-rock miners we sent to Indonesia, we lost about fifteen before they even got to the mine site. Most of these, we discovered, were arrested in San Francisco, Honolulu, or Cairns, Australia for public drunkenness—or something worse. A few seemed to disappear off the face of the earth, and we never heard from them again. Of the two dozen that actually made it to Irian Jaya, ten quit within two weeks; they just didn't like the place. The remainder quickly departed because of problems at the town site, almost always involving some sort of altercation. Within six months, we had only two American miners left.

Long before attrition had reached this point, we saw the need for another approach and sent recruiters to the Philippines. There we hired 50 experienced men from the copper mines who became our training cadre. Not only were the Filipinos quality miners, but they also had an excellent attitude toward their work and got along well with the Indonesians. Our experience with Filipino miners was so good that, between this project and later expansions, we have had at

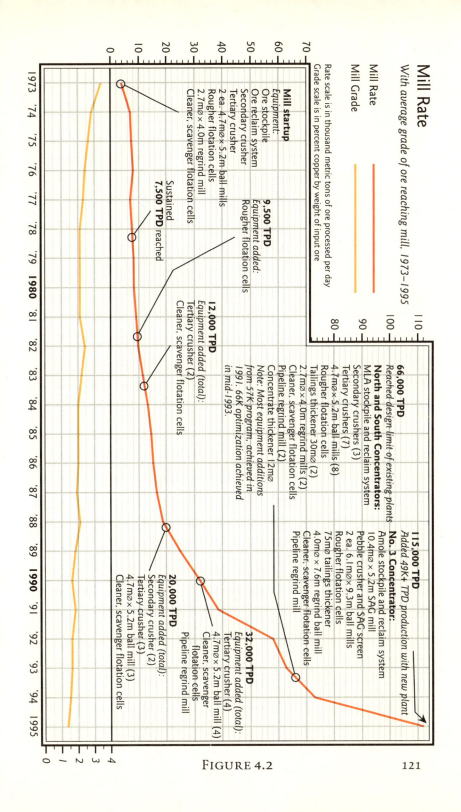

FIGURE 4.2

times up to 330 Filipinos on the payroll—mostly miners, but also mechanics, electricians, and supervisors. Our program for underground mining could not have been accomplished without them.

Underground mining of the GBT deposit began in late 1980, and by 1982 reached its design capacity—4,500 tons of ore a day. Enough ore was coming from the underground mine that open pit production from Ertsberg was scaled back to 5,000 tons a day. The underground project was completed nearly a year ahead of schedule and for $20 million less than the original $101 million estimate. Over time, underground production would increase and the Ertsberg open pit would be phased out, finally ending completely in 1988.

The $81 million Freeport spent on the GBT project proved an excellent investment. Our operations yielded an annual ton of copper for about $900 in capital spent for plant facilities and mine development. This compared quite favorably to copper mines in other parts of the world, which required $1,500–$3,600 of capital investment for every yearly ton of copper. Overall, if copper were to sell for 70 cents a pound, the project would yield a return on capital of 24 percent. As a low-cost producer, GBT would be able to survive periods of low prices, and produce high returns during periods of higher prices. At the time we projected that copper prices would rise and, in fact, 1979 prices averaged 90 cents. In the course of developing the underground mine, we upgraded the tramway system and built dorms, mess halls, schools, and offices in the town of Tembagapura to accommodate the new workers.

Gunung Bijih Timur paid off for Indonesia as well. Mine production increased by 25 percent, and the life of the whole project was extended at least ten years—with a good chance that it would continue much longer. And, very importantly, the training and technology being introduced would outlast the project. Underground mining is a labor-intensive operation, and although by 1982 we had a training group of 410 expatriates, including 311 Filipino miners, more than five times this many Indonesians worked in the overall operation: 2,160 men. This was more than double the 911 who worked for Freeport in 1975.

THE INITIAL GBT mine operations used what is known in the industry as a slusher system of block caving. A "slusher" is a scraper, like a big four-foot-wide hoe, that pulls caved and broken ore from the mine production area to a loading chute, where it drops into a mine car. Although the system worked, it soon became apparent that under our conditions it was not efficient. When it caved, the GBT ore broke into large boulders, which plugged the chutes and draw points, and had to be tediously broken up by hand before they could fit the slusher buckets. (See FIGURE 4.3, page 125.)

As our mine area expanded, the newer areas were developed as a load-haul-dump or LHD operation. An LHD operation relies on special front-end loaders that pick up the ore from the production area and transport it to the loading point. The LHD equipment has six- to eight-foot-wide buckets, and could better handle the large boulders of ore. Our operations, rather than using mine cars, soon went to conveyors for transporting the ore from the mine to the tramway system, where it was then taken to the concentrator. The conveyors greatly reduced our cost, and we continue to use them for transporting ore and waste. Freeport has become a leader in the use of conveyors underground, and today our operations are often visited by representatives of other mining groups interested in this technology.

Although the combination of LHDs and conveyors worked wonderfully, initially we encountered a severe problem with our block cave mine. When we undercut the ore body, the rock just wouldn't cave. Our mine openings were simply not wide enough to allow this ore body to collapse. Since the deposit was relatively narrow, we realized that we would have to widen the span of our undercut out into the waste, thus diluting the ore, in order to force the rock to cave.

We decided to drive a series of adits from the upper elevations into and near the area that we wanted to cave. From these adits, we drilled a series of long holes and loaded these with explosives, to force the cave upwards. This slow, costly work continued for several years and severely limited our production.

Finally, in July 1986, I received a call from Les Acton, who told me the cave finally broke through to the surface. The collapse was spectacular, as millions of tons of waste rock fell into the cave and overflowed down the mountain into the valley below. In Tembagapura, people felt a shock wave and thought it was an earthquake. The fury of this event caught us by surprise, and the rock slide destroyed buildings and expensive equipment that we never thought to protect. From that point on, we did, however, have a block cave mine that performed as it should.

A rock-breaker at work on IOZ ore. Once reduced in size so it will pass through the 18-inch grate, the ore rides on a system of conveyors and down ore passes to the mill.

Despite the problems we were having with the block cave, as we approached a daily production of 9,500 metric tons, the operations became more profitable. As our operating rate went up, our fixed costs associated with infrastructure tended to stay constant, which brought down our overall average unit costs—the classic theory of the economy of scale. The benefits of increasing our production rate were such that in 1981, the mine development department recommended increasing capacity to 12,000 tons a day. This could be accomplished by further modification of the trams and, at the mill, by adding another tertiary crusher and more flotation cells. Helping to justify this increase was the additional 18 million tons of underground ore reserves discovered below the 3,600 meter level.

Achieving a sustained rate of 12,000 tons a day required removing a serious bottleneck—the limited capacity of the trams. We discovered that by reducing the weight of the tram buckets, we could increase their load capacity. The project would also require a new adit, through the wall of the Ertsberg pit from the lowest level, to provide access to the new deep lying ore body. The tram modifications and

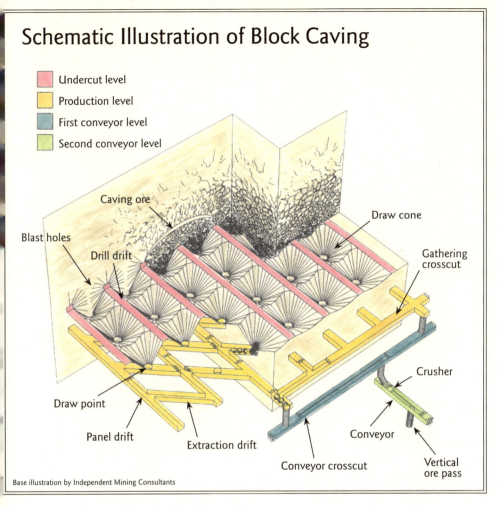

FIGURE 4.3

new adit would cost $47 million. Although copper prices were weak, averaging just 67 cents a pound, the benefits of increased throughput were enough to convince the board, and work on the 12,000 tons a day—called the "12K"—expansion began in 1982.

The project was in full swing in 1983 when we encountered problems with the Indonesian department of manpower. Because of the expanded scale of activities in the underground mine, we needed to hire more expatriate miners. When we appealed to the department to bring in a few more Filipinos, not only did they deny our request, but also advised us that all our Filipino miners would soon lose their work permits. Our mine operators were so frustrated with this situation that they had doubts about the company being able to carry through with the expansion.

Milt Ward and I were asked by our operations managers to meet with them in Hawai'i to review the project. They startled us by recommending that we abandon the expansion. They proposed that the company complete the mining of the higher level ore, shut down the operation, and leave Irian Jaya. If we stopped the development of new areas, and made a small reduction in the number of Filipino miners to appease the department of manpower, they said, we could maintain production and continue to see a profit. After careful review, we turned down their proposal. This strategy was perhaps a good way to exit without losing money, but we weren't ready to leave. We were sure there must be long-term value in this property.

Later that year I found myself in Indonesia making a plea to the director general of the Indonesian department of mines and energy, Lokeito. After a lengthy discussion and several meetings, he suddenly proposed a broader meeting with representatives from all the interested government ministries, where we could explain our need for additional expatriate workers. This was a very surprising turn—Lokeito, who in the past had been a constant obstacle, was now genuinely trying to help. We came back to Indonesia for this meeting, and Lokeito himself came through as a solid and enthusiastic supporter, and made an eloquent and accurate plea for our case. In the end, we were able to keep the Filipinos already working for us, and to bring in the few additional we needed, I believe largely due to Lokeito's efforts. The project continued, and by 1984 we were processing 12,500 tons of ore a day.

Even before we reached our 12,000 tons a day goal, Dick Stewart and the mine development department had conducted some further studies and, in February of 1984, recommended that we expand our production rate to 15,000 tons. The tramway, even with its lighter buckets, was the limiting factor, and they proposed driving an adit, at the 2,900-meter level of the mill, all the way to the GBT ore deposit. The mined ore from the upper levels would be delivered to this tunnel through an ore pass, a vertical shaft down which the rock is dropped, and a conveyor would then take it economically to the mill. Eventually, they reasoned, the ore pass and conveyors could bypass the tramway altogether. The adit would have an additional benefit—from it we could easily prove out the deepest parts of the ore body. This expansion would cost $88 million, included added milling and concentrating capacity, and work was begun in early 1984. By 1985, production had reached 14,500 tons a day, and drilling from the new adit had added another 60 million tons of mineable ore to our reserves.

In 1986, our total reserves stood at approximately 100 million tons and were still growing, which should have been an enviable position. But some of our directors were worried. One thought that our increasing reserves just made us a target for nationalization, or for unwanted interest from one of the growing group of Indonesian tycoons. Other directors worried about low copper prices—in 1985 and 1986, the price of copper averaged just 64 cents and 62 cents a pound—which created severe financial pressures for the company, and a perception that we should conserve cash.

The result of these feelings on the board was a curtailment of most of our exploration programs, which seemed a real tragedy to those of us in the operating arena. In fact, we never quite shut down our exploration program entirely, but instead carried on a "stealth" exploration program. We still needed to drill to define areas of the known deposit that we would soon be mining, and by choosing our locations carefully were able to continue to widen the ore body in areas where it was still undefined. This policy did force our geological staff to do everything on a shoestring, and use salvaged equipment and facilities. Certainly, at this time, no formal budget could have been presented for pure exploration geology.

The year 1986 brought big changes to Freeport. The corporate headquarters moved from New York to New Orleans, and chief executive James Robert Moffett brought a new philosophy to our operation, including a call for all divisions to renew exploration to increase reserves.* Jim Bob Moffett realized that reserves were being depleted throughout the Freeport companies and, if nothing were done, within just 10 years the company would cease to exist. When we at Freeport

---

*I have always suspected that the call for reserves went to P.T. Freeport Indonesia by "accident," just to be consistent with other divisions. Copper was still considered an obsolete commodity, and the company's interest was really still in gold, sulfur and fertilizer. Nevertheless, those of us who believed in the Irian Jaya operations didn't bother to question why the call went out, we just acted on it.

One of the special loaders used in the GBT load-haul-dump operations deposits its ore in a truck. The LHD process proved more economical than the previous slushers.

Indonesia received the call, we did not need much encouragement, and began exploring out in the open once again.

By June 1987 we had added sufficiently to our known reserves that we were able to recommend increasing production once again, this time to 20,000 tons of ore a day. This project, soon called simply the "20K project," would cost $158 million, and required an expansion of the mill, including a new concentrator building, as well as improvements to the mine facilities and the town of Tembagapura.

We also needed more power, which we recommended be accomplished by building an underground hydroelectric plant. The plant would be installed just off to the side of the Mill Level Adit, at the elevation of the concentrator. It would receive its water through a vertical, 500-meter-long, 1.5-meter diameter passageway that extended through solid rock up to the elevation of the mine. The rock was of such good quality that the aqueduct required no lining or grouting.

By August the board approved our plan, and construction began immediately. The operation reached a sustained 20,000 tons a day in 1988. With the revenues this was bringing and our continuing explo-

Workers at the underground mine enjoy a break. An initial cadre of skilled Filipino copper miners was crucial to our success with the GBT block cave mine.

ration success, many of us were saying, not too quietly, that we would be in Irian Jaya for another 100 years. Gross revenues had increased from $74 million in 1975 to $400 million in 1988. While the number of expatriates on the job remained about the same, 400–410 men, Indonesian employment had reached 3,450.

The expansion required the support of both banks and customers, and both were quickly forthcoming. At this time, Freeport Indonesia's concentrate was beginning to make a name for itself in the market place as a premium product. Our concentrate is remarkably "clean," containing very little arsenic, antimony, bismuth and fluorine—elements which either cause environmental problems at the smelter or reduce the quality of copper metal produced.

One particular phase of this project deserves special mention— the ore passes. As part of the 20K project, we completed three ore passes, each 550-meters long and 2 meters in diameter. From the 15,000-ton bins beneath the ore passes, conveyors delivered ore to the concentrator. An ore pass is like a laundry chute, except that instead of dirty clothing, chunks of raw ore are poured into the top from the end of a conveyor belt. Needless to say, the impact of rocks against each other after falling into an ore bin half a kilometer below results in a considerable reduction in their size. The crushers and grinding mills thus receive much finer rock, and have much less work to do. Most importantly, however, the ore passes circumvented the tramway, which was the bottleneck of the whole system. Our trams could carry only 18,000 tons of ore a day, and even if we were willing to pay the price, there was simple no space in the narrow gorge to add more, even if we were willing to pay the price. An ore pass, on the other hand, can transport as much ore as you can pour down it. Being able to bring large quantities of ore from the mining areas to the concentrator lowered our costs, and set the stage for even higher levels of production.

The operating success of this system rivals the discovery of the Grasberg in its importance. Grasberg could not have been developed to the degree we see today without the success of the ore passes. Freeport's use of ore passes is unique. The El Teniente copper mine in Chile moves only about half the volume of ore through ore passes as we do, and only Freeport allows the rock to free-fall as far as 550 meters down the center line of an ore pass.

In retrospect, the 20K project, with its facilities and reserves, was the key to the recent growth of Freeport Indonesia, and to the formation of the entity we see today, Freeport-McMoRan Copper & Gold. Despite having mined 54 million tons of ore in the previous 14 years, the year 1987 found us with 88 million tons of ore still to mine. Drilling was actively taking place in the IOZ and the DOZ, and the reserve picture was rapidly expanding. Freeport's engineers were certain that with the completion of ongoing projects, our ore reserve would expand to at least 133 million tons from known geological resources, and by drawing our ore from multiple sources, we greatly reduced the risk of costly interruptions in production. We were certain we could expand our reserves by exploring down dip and along strike from existing ore bodies—basically, following the fault system—and by exploring other known targets in the COW area.

We were now benefiting from the economies of scale associated with higher levels of production. If copper prices were to average 80 cents a pound, we would receive a 40 percent return on our investment. The previous five years had not been good for copper producers, however, because prices had stalled in the 60–70 cents range. But even at an average of 62 cents a pound, our 20K expansion would yield a return of 22 percent.

Five years of low copper prices had taken a toll in Freeport's boardroom, and managers and board members began to look negatively on copper operations. This was not helped by the spate of articles in the popular press of the mid-1980s touting the benefits of fiber optic cable over copper wire for communications, and the benefits of aluminum over copper for car radiators and home wiring. Amazingly, these articles claimed, the copper mine of the future lay under the streets of New York City, where copper telephone wires would be torn out and replaced with fiber optic cable. Today, despite the popularity of fiber optic cable and aluminum radiators, copper remains a heavily used and even fast-growing commodity. Aluminum has proved an unsafe replacement for copper in home wiring, and an unsuitable replacement for many industrial uses such as motor windings. But in the mid-1980s, these negative arguments were persuasive.

As a result, several attempts were made to sell off either part or all of Freeport Indonesia, to obtain some value for the mother company, Freeport-McMoRan. For those in the company championing the copper business and Freeport Indonesia, this seemed to be a mistake, as even during the previous lean years, Freeport Indonesia had paid out dividends of $10–$15 million a year. With our efficient operations and high rate of production, all it would take would be a rise in copper prices and the cash dividends would be both significant and immediate. The directors of the mother company commissioned a financial evaluation of Freeport Indonesia, which concluded that its value was no more than $200 million. If someone would buy Freeport Indonesia for this price, the mother company could put the capital to use in more profitable "growth" areas.

At the same time copper fell out of favor in the boardroom, interest in gold rose dramatically. Freeport, in fact, seemed to be converting itself to a gold company, entering at every opportunity into gold exploration projects in the United States, Australia, and South America. Our explorationists in South America were pointedly told not to bother with any copper or other base metal property, just to concentrate on gold. With gold hovering around $450 an ounce, the price earnings ratio of a gold operation, as expressed in a stock price, was a siren call. Gold fever swept the board room, and our copper operation in Irian Jaya was of little interest in New Orleans.

It was also, thankfully, of little interest to outsiders, and no sale took place. The rest of the world seemed to have the same attitude regarding the copper business, and there were no buyers at any price. It was difficult to convince potential buyers to even visit the property. In 1987, Freeport-McMoRan decided to demonstrate the value of its asset, P.T. Freeport Indonesia—and at the same time obtain some cash value for it—by undertaking a stock offering on the London Metal Exchange. London was picked because it was considered a better market than New York for an Indonesia-based copper issue, and we began working with British financial advisors to prepare the public offering.

In October 1987 the world stock markets suffered a sudden downward correction of more than 25 percent, and the planned offering was withdrawn. If this downturn had not occurred, the company would have been listed in London instead of New York. In October 1988, by which time the world's stock markets had recovered, the effort was reinitiated. This time the stock was offered on the New York Stock Exchange, under the guidance of Kidder Peabody. The offering, well oversubscribed, was a grand success.

ONE SMALL EVENT associated with the abortive London public offering unexpectedly turned out to have momentous implications for the future of the company. At one point in the process, in describing our assets, I was asked by our advisors if there were any gold potential at our location. The London financial community also seemed to have gold fever and thought that if we could add a little "pizzazz" to the description, the issue would sell better. I was able to say, truthfully, "Yes, there certainly is some gold potential."

On a trip to Irian Jaya, I asked one of our mine geologists if he had come up with anything interesting lately. This was Dave Potter, at the time assigned to the mine operation from the Freeport Gold Company exploration group in the United States. We then had a practice of moving promising exploration geologists to the geology department of an operating mine for a period of time to broaden their knowledge. Potter said he had picked up a few samples from the top of Grasberg, and when he had them assayed, they turned out to be "strongly anomalous," as geologists say, bearing 1–2 grams of gold per ton. These are very good numbers, and even better, this "anomaly" seemed to extend over a very large surface area.

Ertsberg production waned through the 1980s, and the mine was decommissioned in 1988. Today the pit is Lake Wilson, and stores water for the mill.

Potter, now vice president for exploration, was one of the few geologists at the time who had any interest in Grasberg. It was almost universally considered a low-grade copper porphyry that was not even worth exploring, much less mining, and our senior geologists thought it a waste of time for any of their geologists to be walking around up there. In fact, Potter collected the samples that turned up the high gold values on his own time.

I first described the Grasberg deposit to our financial advisors as a "copper-gold porphyry," as we expected to find both metals, as well as some silver, at this location. I also noted that the geology of Grasberg looked exactly like that of Ok Tedi, a copper-gold porphyry being mined in Papua New Guinea, which a group of us from Freeport had then recently visited. Ok Tedi received considerable publicity at the time because of the deposit's rich surface capping of gold. This analogy further intrigued our "gold bugs." But copper was still so maligned in investment circles that our advisors first suggested that the very mention of the word "copper" would turn away potential investors. Eventually, they agreed that we could keep "copper," as long as we put the word "gold" first. Thus Grasberg became a potential "gold-copper porphyry."

If we mentioned Grasberg as an exploration prospect, our advisors noted, we would have to, in good faith, commit to a definite exploration program on this deposit. I was more than happy to oblige them. We were all still chafing from the hiatus of funded exploration. We soon announced a modest five-hole exploration program, and began diamond drilling Grasberg, which could have given our senior geologists heartburn. Although the October stock market correction stopped the London offering, the Grasberg drill program continued. A promise was a promise.

By early spring of 1988, we had several drill holes into the near surface areas of the Grasberg stockwork zone. As the holes got deeper, we continued to encounter good gold values, but found ourselves facing increasingly interesting numbers for copper. Finally, on June 4, 1988, the last hole of the program, drilled 611 meters straight down, was completed. This drill hole contained 591 lineal meters of 1.69 percent copper and 1.77 grams per ton of gold, probably the single most spectacular core sample ever produced in the copper mining business.

I learned of the data on this hole in New York during one of a series of formal presentations during a road show to present our stock offering to the investing public. Jim Bob Moffett had just finished a presentation, and I stepped aside to check my messages with my secretary. She read me the figures over the phone. Our instructions from our advisors during the course of presenting the stock offering were not to provide any updates to the information already provided in writing, and particularly to remain absolutely silent on any news from on-going drilling programs. As the importance of what was being read to me began to sink in, my eyes wandered across the room and saw Mark Cohen, a very astute analyst for Kidder Peabody, keeping a close watch on me while feigning interest in a conversation he was having. One of the most difficult feats of self-control I have ever accomplished was to thank my secretary, hang up the phone, and walk back to the table as blithely as if my wife had just told me what she was planning for dinner.

Although I could not yet prove it, my gut feeling was that Grasberg was going to be a world-class deposit, and Freeport-McMoRan Copper & Gold was going to have a fabulous future. There were, and still will be, plenty of challenges. But the Grasberg deposit, as we know it today after seven years of continuous drilling, contains the largest known reserves of gold of any mine in the world, and the third-largest known reserves of copper. This has turned out to be one of the most important mineral discoveries of the century.

The Grasberg mine, with Puncak Jaya in the foreground. This deposit contains the largest gold reserve and third-largest copper reserve of any mine in the world.

CHAPTER FIVE

The Grasberg Era

A mistaken assumption—The porphyry model—A disappointing test hole—Gold—Rapid expansion—Conveyors and ore passes—The black chicken—An ancient volcano—Block A riches—Exploration challenges

GRASBERG IS BY far the biggest and brightest jewel in Freeport's crown. As of early 1996, the deposit has been found to contain 1.76 billion metric tons of ore, with an average grade of 1.11 percent, or 35.2 billion pounds of recoverable copper metal. The gold content is astoundingly high, 49 million troy ounces, almost half of all the gold extracted during the California gold rush. Yet this huge deposit, less than three kilometers from and within sight of Ertsberg, was not even drilled until 15 years after work on the original mine began.

Conventional geological wisdom suggested that Grasberg, which had been noted and named by Jean Jacques Dozy more than 50 years before, was at best a low-grade copper porphyry. Sometimes these deposits will have a "cap" of enriched copper and gold, but if this had ever existed on Grasberg, our senior geologists claimed, it was scraped away by the recent glaciers. If there were any copper ore left in Grasberg it would be of very low grade, or even just "protore," unenriched igneous rock with mere traces of copper.

Throughout the 1980s, geologist Dave Potter stared across the Carstenszweide at Grasberg from the run-down former welding shack that then served as the mine geology offices. Potter's background was not in copper deposits; before coming to Irian Jaya he explored disseminated gold deposits in the Great Basin district of the United States. The high gold values in samples from Grasberg interested him, and he was intrigued by what the mountain might contain.

"My 'eye' was set for gold," Potter says. "In retrospect this was probably a good thing, as I was not experienced enough in copper porphyry systems to know that there was nothing there!"

Potter was not convinced that glaciation had recently scoured

Grasberg, or that the generally poor surface showings of mineralization could be taken as indicative of the mineralization inside the mountain. The surface rock, he thought, could be generally low-grade because of weathering. I am not a geologist, but I was raised in Alaska and have seen a lot of glacial topography. I agreed with Potter that the top of Grasberg must have stuck above whatever glaciers may once have covered the area, or the topography would look quite different. By the late 1980s, Potter and I, with support from chief operations officer Milt Ward, had become interested in putting some holes into Grasberg.

Milt had always felt Grasberg had something more going for it than others thought. He asked us all many times if we were satisfied with what everyone said about Grasberg—maybe we should look at it again? He, out of everyone, was positive about Grasberg's potential during the early, crucial stage.

A drilling program is not cheap, however, and our senior geologists would never allow so much money to be spent on what they saw as a futile effort. It was the interest of the financial community in gold, and our need to back up our claims to having a gold prospect by embarking on a real exploration program, that finally gave us the opportunity to drill Grasberg.

Grasberg looked nothing like the other peaks standing around it. It was low enough to have been covered by trees, but instead the only vegetation that grew on it was a type of coarse grass. This "vegetation anomaly" is one of the clues that field geologists look for. In the case of Grasberg, the growth of trees and brush was inhibited by the acidic soil—tolerated by the hardy grass—which was the result of all the copper- and gold-containing sulfides being leached from the rock as a result of natural weathering.

This, of course, is what we know now. There are many vegetation anomalies in the world that have nothing to do with commercial mineralization. For example, the bald, grassy hills west of Nabire, a town in Irian Jaya north of the Freeport area, result from a bed of quartzite, containing no copper, gold or sulfide minerals whatever. Quartzite has nothing but silica in it, and it is the nutrient-poor character of quartzite-based soil, not its acidity, that limits the vegetation there to tough, rangy grass.

Any geologist would be intrigued by Grasberg's appearance, and knowing that there are intrusive systems in the area, might at least hypothesize that it was the result of mineralization. But there are many, many expensive steps between a field geologist's hypothesis and a proven ore body. A program of deep diamond drilling, with visual and chemical evaluation of the cores, is required to "prove" a deposit. And even if metals are found, a grade that may be economical to mine in a dry, accessible, low-altitude site in the United States may not be worth the cost of extraction in the highlands of Irian Jaya.

ALTHOUGH HE WAS the first to sink a test hole into the mountain, Potter was not the first geologist to show an interest in Grasberg. The first was Jacques Dozy, who named the mountain during the Carstensz Expedition of 1936. "The rather smooth grass-covered Grasbergen," he writes in the 1939 report on the expedition, "form a striking morphological element amidst the limestone mountains." Dozy also noted the presence of iron bogs and sulfide minerals in the area. Dutch geologist W. J. Jong analyzed Dozy's rock samples, and in several, from various locations on Grasberg, found signs of mild metamorphism. "Whether metamorphism is due to the porphyritic intrusion of the Carstenszweide, or to a hypothetical, later intrusion... must remain an open question," he writes.

Of the samples of ore Dozy brought back, only three were analyzed, by Dr. Ir. C. Schouten of Delft University. Two came from Ertsberg or nearby, and one from Grasberg. This latter, which was collected on the far western edge of the Carstenszweide, was so weathered and porous that Dr. Schouten had to boil it in a mixture of shellac and resin before he could prepare a polished section for examination. Because the sulfides had been leached from this specimen, Schouten writes, "It could not be ascertained therefore whether... copper ores were present in the primary ore."

The first Freeport geologist to think seriously about the potential of Grasberg was Bob Stewart, our first resident geologist. Stewart was in charge of geology at the fledgling Ertsberg mine, as well as exploration within the 10-kilometer by 10-kilometer contract of work area, or COW, surrounding it.

"Stewart had plenty to do in the mine itself," recalls geologist Frank Nelson, who was brought to Irian Jaya in 1973 to help explore the COW area. "In the beginning at least, he went to the field to do mine geology plus the exploration within the COW, but it became obvious that the mine geology itself was almost more than anybody

Grasberg in 1988, before mine development began. At left is the Carstenszweide, stretching back to the Ertsberg pit, and in the background right is Lake Wanagong.

could handle. He did work up on Grasberg and in other places, but it was impossible for him to spend much time there."

Stewart was an old Freeport hand who had come to Irian Jaya from the company's exploration office in Tucson, Arizona. The copper deposits found in the American Southwest are porphyry systems, like Grasberg, and Stewart had spent a lot of time looking at the rocks associated with these ore systems in an environment where the lack of vegetation made the surrounding alteration very visible. Stewart's experience "set his eye" for recognizing potential porphyry systems, and he had the curiosity of a true explorationist.

"Bob was in charge of the mine geology of the new Ertsberg open pit mine, but he just couldn't spend his time looking in, he had to look out," Potter said. "So he was messing around where he wasn't supposed to be, doing what he wasn't supposed to do, and collected a whole bunch of data, samples, some maps and some thoughts on the Grasberg that he put into a report."

So even as early as the mid-1970s, Grasberg had been suggested as a potential exploration target, backed up by Stewart's report. But

Now one of the highest-production mines in the world, Grasberg, though just a few kilometers away, was not drilled until almost 15 years after the Ertsberg mine opened.

no drill rig touched Grasberg for another ten years. "The 'experts' just knew that there was nothing there of value," says Potter, with some frustration. "Stewart suggested a drill hole but got talked out of it in the face of the overwhelming 'evidence' and experience of others." This guaranteed the report a place on a dusty shelf in Reno, Nevada, where Potter found it years later.

"Where Bob went wrong, in my opinion," Potter says, "was not to have pushed for a drill hole, and not to have considered the action of a tropical environment on the copper ion—it disappears into the rivers very fast."

There was some justification to the Freeport management's reluctance to investigate Grasberg. At that time theories of porphyry copper systems were based on the southwestern United States model, large deposits with an average primary copper grade of less than .5 percent. It was, in fact, a defining characteristic of porphyry systems that they be formed from low-grade primary deposits with some secondary enrichment. The notion of a high-grade—over 1 percent copper—so-called "Pacific model" for a porphyry system did not begin to appear until the mid-1980s. In other words, in the early 1970s there was no available analogy that would have allowed for the existence of such a large, high-grade porphyry system.

It is important to remember that, for a mining company, "ore" is an economic category, not a geological one. To a geologist, a few hundred parts per million of copper is a geochemical anomaly, and a few tenths of a percent is interesting mineralization. But at what point it becomes "ore" depends on how much it costs to extract and concentrate. To the best of anyone's knowledge at the time of Bob's effort, the best grade the Grasberg porphyry could possess would be about .3 percent. And at that place and at that point in time, this certainly wouldn't have met the definition of "ore."

Even as skilled and respected a field geologist as Frank Nelson didn't question the traditional thinking on Grasberg. "Nobody dreamed of how rich it was going to be," he said. "Everybody assumed that it would be a normal—quote low-grade unquote—ore, and what are you going to do with it up there at twelve thousand feet in the middle of the New Guinea jungle? I guess that teaches one not to assume and to go ahead and drill."

Grasberg's problem, according to the traditional porphyry model, was that it lacked what is called a "chalcocite cap." This cap is a characteristic of desert porphyry systems such as those found in Arizona, Chile and Peru, where the original ore is enriched through weathering. Copper sulfides are leached from the chalcopyrite and redeposited in a surface-stable form of copper called chalcocite. In this way, the grade in the surface layer—sometimes tens of meters thick—can be improved by several times over the original. It was believed until recently that this chalcocite cap, if it ever existed on Grasberg, was stripped away by the glacier, leaving only the low-grade primary mineralization, or protore. The copper values on the surface were very low, a mere .05–.10 percent, which helped convince everybody that their Arizona analogy was correct.

According to Del Flint, who accompanied Forbes Wilson on the original Ertsberg expedition and directed the early '70s exploration program from the office in Reno, Nevada, his geologists were under particular pressure to find high-grade ores. The company needed high-grade ore to get their investment back, and if it wasn't over 1.5 percent copper, they had to leave it in the ground.

It is also important to remember that, although the scale of Grasberg tends to dwarf everything that has come before it, from the 1970s through the mid-1980s the company was adding considerable reserves from the high-grade skarn deposits east of Ertsberg. Although skarns are always smaller than porphyry deposits, the GBT is a world-class skarn deposit. At early 1980s metals prices, the value of the ore in the GBT complex was more than $4 billion, a figure that in today's dollars would be closer to $7 billion. This only looks small because of the massive size of Grasberg. Before mining began, the GBT, the IOZ and the DOZ taken together would have held 100–110 million metric tons of ore, with an average grade of around 2 percent copper. This is a pretty healthy reserve.

The importance of GBT in particular is often overlooked in the Grasberg era. As mining consultant John Marek has said, "The GBT should be given a 21-gun salute and a Blue Angels fly-by." Even just two years ago, during a disruption of one of the Grasberg ore passes, the GBT mine was brought back into 20,000 metric tons a day production to make up the difference. This was from a mine that was almost exhausted. It's a pretty rare ore body that you can crank back up like this in its closing days.

When Potter first came to Irian Jaya in 1983, conditions for geologists were very different from those today. At that time, the mine operations department was concentrating on moving the ore, and exploration geology was not a high priority. Copper prices had dropped 30 percent since 1980, hovering at around 70 cents a pound, and senior management saw no reason to spend money to find more copper. Any available cash would be spent on gold exploration, and all copper exploration would be limited to what was needed to define the ore bodies currently being mined. It looked like we would finish mining Ertsberg and GBT, and get out.

In 1983, the geology office was not even at the mine, it was down by the mill, a carryover from an aborted gold exploration program on the upper Wanagong drainage, which was easier to reach by helicopter from the mill. Nor was the office particularly fancy, Potter recalls:

"Our building was an old tin shed, full of core, rats and mold. It was a little lower than the surrounding area, and when the tailings line backed up, which would happen two or three times a year, the building would end up with twenty or thirty centimeters of soupy tailings throughout. This was hell on our paper maps and files."

Potter eventually got new quarters for his department, close to the mine. Still, budgets were tight, and he was given an old structure that had been a welding shop at the far end of the mine office complex. The $60,000 it would have cost to furnish properly was considered too expensive, so his department scrounged for desks, cabinets and fixtures. But at least from there his men could conveniently get to the mine to do their work. And from the new office, Potter could actually see Grasberg.

Every day the sun came up at the mine, he could look at that mountain towering over the top of the geology office. The alteration was beautiful—to a geologist. Potter spent a number of years watching that mountain, collecting information, walking out of the old welding building and across the Carstenszweide to the base of it and, finally, as the reserve picture increased and his department's budget grew, he could take helicopters up to the top.

Potter drilled his first hole in Grasberg in 1985. There was a bulldozer doing some work out in the Carstenszweide, and he located an old Longyear 24 drill rig that was gathering dust and got a dozer driver to haul it to the edge of the mountain, near a catchment pond. The Longyear 24 has a very limited capacity, and of course he was not drilling down from the top of the mountain, but sideways in from the Carstenszweide. We had no authorization for this exercise, either, so we called it a "training hole."

The small drill could only drive a hole 200 meters in, which was not enough to hit rock that could legitimately be called "ore." The bit intersected intrusive rock that tested at .3–.5 percent copper. Although in hindsight this proved the existence of a somewhat higher-grade porphyry than was then thought possible, the low numbers were a disappointment. At first the hole was dubbed GRS-1—GRS for Grasberg—but when management later saw that the first hole into Grasberg was drilled from the side, and had yielded unimpressive results at that, the hole was renamed SK-1, for South Kali, a nearby stream. The GRS tag was reserved for the impressive holes later drilled into different rock from the top of Grasberg.

Around 1986, Potter saw a water valve that was kept as a souvenir by a maintenance supervisor. The valve came from one of the water supply reservoirs in the mine area, just beneath Grasberg. The reservoirs, which help supply the mill at times of low rainfall, have a tendency to fill with dirt and have to be cleaned regularly. The valve on the supervisor's desk had been naturally plated with copper. Some-

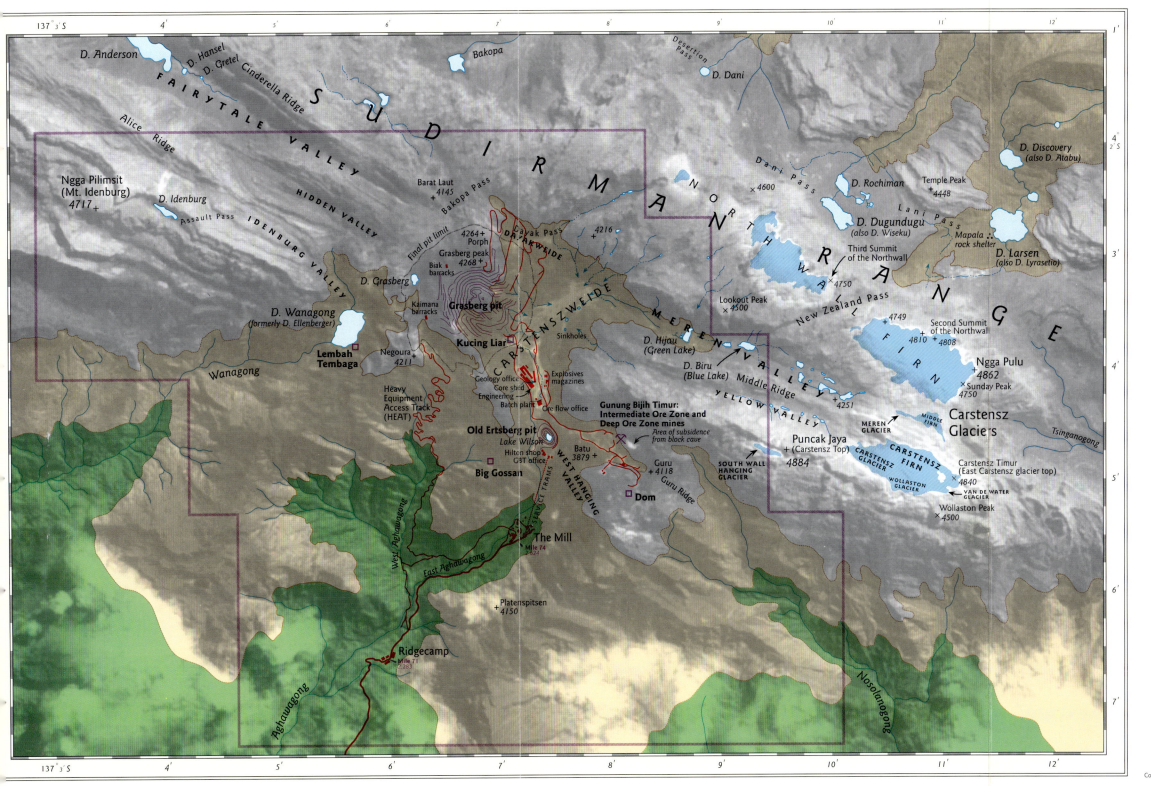

The Grasberg mine is in one of nature's most impressive settings. The long lens used in this stunning shot artificially compresses the distance from the mine to the glaciers.

how, copper was getting into the supply reservoir, and the only place it could be coming from was Grasberg. Despite this interesting valve, and the copper values in SK-1, Potter still wasn't thinking of Grasberg as a copper prospect. His eyes were tuned for gold.

What really began to get Potter excited about the potential of Grasberg were some surface samples that he and geologist Tom Collinson took in 1987. By that point the geology department had experienced a few successes, particularly in extending the DOZ and Dom reserves. They still didn't have much of a budget, however, and with helicopter time running $1,500–$1,800 an hour, Potter and Collinson had to hitch their helicopter rides to the top of Grasberg and back. Potter had to spend a considerable portion of his time in the office and the underground mine, where his primary responsibilities lay, and he called these trips to Grasberg "occupational therapy." The samples he was bringing back, if there were enough tons up there, bordered on ore values for gold. With our upcoming stock offering providing the justification, the assay data from these samples allowed us to begin a proper drilling program on Grasberg.

The top of Grasberg reached almost 4,300 meters, and the Carstenszweide, where the equipment and mine offices are, is 3,650 meters. At these elevations, even our most powerful helicopters can lift only 1,400 pounds. Potter wanted to be able to reach depths of more than 500 meters, and this would require a larger capacity rig than the little Longyear 24 he used at the base of the mountain. Fred Elliott, one of our most experienced drillers, was put in charge of the project in late 1987.

"Some correspondence with the Caterpillar people in Australia at that time gave us an idea of what we were up against," Elliott says. "I was trying to spec an engine to give us adequate horsepower at the forty-one hundred meter elevation. Their response was, 'Since human beings can't be expected to work at that altitude, how could I expect an engine to work?' So much for further correspondence with the Cat people."

Elliott identified a couple of drills with the proper capacity, but these had been used in underground work, and were configured with electric motors. Of course there was no electricity on top of Grasberg then. Leon Greenfield, then our diamond drilling instructor and a master bush mechanic, had the job of organizing the first rig. There was no time to get new rigs in, so the available ones would have to get new power plants.

"We ended up having to 'borrow' two used Toyota diesel truck engines to power the rig," Elliott said. "Leon was able to get one of them rebuilt in time, but the lack of time and money meant the other had to limp along the best it could. The drill was powered by hydrostatics, so the best engine ran the primary pump, and the other ran the secondary pumps and the drilling fluids pump."

When our drilling program began, our working theory was that the first 50 meters or so would have the highest concentration of gold because of the weathering of the surface rock. This pattern is often seen in wet, tropical climates. Thus the first hole, started—or "collared"—on January 28, 1988, was positioned so that it would nearly parallel the surface of the ground, at a depth of only about 25 meters.

I was very pleasantly surprised when Potter telexed the initial assays from this hole in late February. There was gold, although not the concentration we had expected. The hole did, however, intersect three zones of copper which could be considered ore grade.

I picked up the phone and called Milt Ward. I told him I had some good news, and I had some bad news. The good news was that we had encountered a total of 144 meters of 1.69 grams per ton of gold. The bad news was that we had encountered 144 meters of 1.63 percent copper. He and I both wondered whether the presence of all this ore-grade copper could be significant. It certainly deserved a closer look, and we both anxiously awaited the results of the next holes. Potter drilled three more holes, at progressively steeper inclinations. The results came in better still, and our excitement grew. Finally, we drilled the fifth and final hole of the initial program, 611 meters straight down. This hole, GRS-4—the holes were drilled out of order—intercepted ore for almost its entire length, hitting 591 meters of 1.69 percent copper and 1.77 grams per ton of gold. We knew from this that we were not dealing with some surface phenomenon. This was something deep-seated. (See TABLE 5.1 on page 148.)

"GRS-4 was really the hole that turned the tide of the project," Potter says. "Up till then, everyone was convinced that it was a fluke, a smallish enriched area along a fault, or something that was not indicative of the area as a whole." Actually it would take several dozen more holes to convince everyone that we had a major porphyry system in that mountain.

This project certainly warranted the full support of our resources. We began mobilizing additional drills, robbing from the underground mine, and contracted a Bell 212 helicopter to devote exclusively to the Grasberg drilling program. Excitement increased as the results came in. By February, 1989, 47 drill holes had been completed, yielding 13,929 meters of core. Drilling was still continuing, but by early 1989 our geologists and engineers estimated that Grasberg contained at least 99 million metric tons of ore grade rock.

WE QUICKLY DEVELOPED a pit plan for mining this material, and although we knew the deposit would continue to grow, we produced a preliminary feasibility study which became known as the "32K Program"—Phase 1 of the Grasberg project. The study suggested that the capacity of the mining and concentrating facilities could be increased from the current 22,000 tons per day to 32,000 tons per day by the middle of 1990. We also began studying the possibility of further expansions, so that we would be prepared if our geologists found more reserves—as we fully expected they would. The 32K program was estimated to cost $125 million.

Grasberg Drill Hole Timing and Results
Data from George D. MacDonald, 1995

The initial program was designed as five holes laid out along a narrow ridge, directly above the stockwork zone. Drill hole numbers were assigned according to the original drilling plan, but were drilled out of order. Hence, the hole number sequence is not chronological. In the following tabulations of ore intercepts, intervals are defined using a .70 % Cu cutoff. A total of the ore intervals within each hole is also provided.

GRS-1. Collared April 9, 1988 at -45 degrees inclination; completed to 607 meters May 9, 1988.

Hole I.D.	From (m)	To (m)	Length (m)	% Cu	ppm Au	ppm Ag
GRS-1	24	284	260	1.22	1.28	2.27
GRS-1	294	314	20	.76	.72	1.13
GRS-1	418	450	32	1.35	1.14	

3 zones, totaling 312 meters, averaging 1.21 % Cu, 1.23 ppm Au, 2.18 ppm Ag

GRS-2. Collared January 28, 1988 at -15 degrees inclination; completed to 321 meters February 20, 1988.

Hole I.D.	From (m)	To (m)	Length (m)	% Cu	ppm Au	ppm Ag
GRS-2	54	166	112	1.73	1.70	3.46
GRS-2	200	216	16	1.30	1.61	1.00
GRS-2	258	274	16	1.29	1.73	2.25

3 zones, totaling 144 meters, averaging 1.63 % Cu, 1.69 ppm Au, 3.06 ppm Ag

GRS-3. Collared February 22, 1988 at -25 degrees inclination; completed to 593 meters April 5, 1988.

Hole I.D.	From (m)	To (m)	Length (m)	% Cu	ppm Au	ppm Ag
GRS-3	78	220	142	1.44	1.31	1.94
GRS-3	238	296	58	1.41	2.07	3.62
GRS-3	304	354	50	1.60	1.99	3.80
GRS-3	382	468	86	1.12	1.17	3.21
GRS-3	474	593.2	119.2	2.05	1.50	7.04

5 zones, totaling 455.2 meters, averaging 1.56 % Cu, 1.51 ppm Au, 3.94 ppm Ag

GRS-4. Collared May 20, 1988 at -90 degrees inclination; completed to 611 meters June 4, 1988.

Hole I.D.	From (m)	To (m)	Length (m)	% Cu	ppm Au	ppm Ag
GRS-4	10	80	70	1.46	0.45	1.90
GRS-4	90	611.2	521.2	1.72	1.95	2.59

2 zones, totaling 591.2 meters, averaging 1.69 % Cu, 1.77 ppm Au, 2.51 ppm Ag

GRS-5. Collared May 16, 1988 at -90 degrees inclination; completed to 206 meters May 27, 1988. (Note: This hole was drilled from a new site using a second drill.)

Hole I.D.	From (m)	To (m)	Length (m)	% Cu	ppm Au	ppm Ag
GRS-5	42	205.8	163.8	2.40	3.64	4.20

1 zone, totaling 163.8 meters, averaging 2.40 % Cu, 3.64 ppm Au, 4.20 ppm Ag

TABLE 5.1

The first drill rig on Grasberg, in spring 1988. The rig is drilling GRS-3, a long, shallow hole that attempted to follow the predicted richest zone of mineralization.

The mining strategy called for mining 20,000 tons from the underground mines and, at first, 12,000 tons from Grasberg. As a pioneering project, the Grasberg mine strategy would rely on front-end loaders and relatively small 85-ton trucks. Under the initial conditions we envisioned at the Grasberg mine, we considered the smaller and more maneuverable equipment to be more appropriate than the giant, high-volume machinery working today.

The underground ore would be delivered to the mill by the ore cars on the tramway system and the ore passes. The 12,000 tons per day from Grasberg would be delivered on a new conveyor system that would bring the ore from the new pit to the existing underground ore pass and conveyor system, which would be upgraded to handle the increased capacity. To increase mill capacity, a fourth primary ball mill was added, and the tertiary crusher and rougher flotation areas were expanded. We also had to upgrade the power plant, and add infrastructure to Tembagapura to accommodate the new workers we would need.

The feasibility study was reviewed by Freeport Indonesia management as well as the top managers of Freeport-McMoRan Copper & Gold and Freeport-McMoRan in New Orleans. Pending final board approval, we were given permission to place orders for major equipment and begin construction. This was not going to be easy. We didn't even have a proper road across the Carstenszweide to the base of Grasberg, much less a road to the top. We would need to install crushers and ore passes and a conveyor system.

There was considerable debate in the early days as to the best design of the ore delivery system. One option was to locate the primary crushers off to the side of Grasberg, and from there to deliver the ore to the existing ore passes used by the underground mining operations. The other option was to locate the primary crushers on the top of Grasberg, and to drop the ore through ore passes to a conveyor system running along an adit through the mountain to the existing underground mine conveyor system.

We elected to go with the second of these options, because it would allow us to bring the mine into production sooner. We would continue to drill to evaluate the size and limits of the Grasberg deposit, and later, when we knew better the shape of the deposit, locate a permanent crusher off to the side.

This meant, however, that we would have to mobilize underground miners to drive an adit almost a kilometer long into the side of Grasberg, and to drill and ream 550-meter ore passes. We had the resources, but all of this work would have to be completed in record time. Our people rose to the occasion.

This construction was a painful process, because the rock in the area of the portal was unstable, and steel rods, called "spilling," had to be driven into the soft rock to provide support. Initially our crews were having a horrible time with this, advancing only inches a day. This adit was critical to all the other aspects of the project, and this delay was putting the entire investment in jeopardy.

The Filipino supervisor of the project came up with a unique solution. He brought a chicken from the town site, a black chicken, and cut its head off and dribbled the blood around the site. The next day the crew was able to proceed with the spilling, and quickly reached more stable rock. We never again had a problem with this adit. The underground crews began to advance faster than scheduled. From that day forward the adit was named "Ayam Hitam," which means "black chicken" in Indonesian. The Indonesians have picked up on this technique, and I am told that whenever there is a serious problem of this type, a black chicken is sacrificed. Perhaps the technique should be added to the next edition of the *Mining Engineers Handbook*.

As early as May 1989 a more definitive study had been completed outlining the company's intermediate term plans for the Grasberg. We could see from the core data coming in that in all probability, Grasberg would be a very large ore body. Further, it would probably take several years to ascertain the full size of this deposit. We did not want to wait this long, so we looked at the concentrator site to see how we could add capacity to the facilities.

We determined that we could add three additional primary grinding mills without major and costly site preparation expenditures, which would bring the plant to a total throughput of 57,000 tons per day. This was Phase II of our Grasberg expansion, the 57K Program. The cost of moving from 32,000 tons per day to 57,000 tons per day would reach $381 million, and as of May 1989, construction was not yet complete on some of the infrastructure additions of the 20K Program. Including remaining 20K and 32K work, and some already scheduled improvements, our operations and engineering group was now facing $716 million worth of capital programs to complete over the next four years.

The 57K program involved the demolition of several buildings on the south side of the original concentrator, which we used for storage and maintenance purposes. This space would be used for what came to be called the South Mill Addition, to which the 57K expansion contributed three additional primary ball mills and entirely new tertiary crushing and fine ore storage bins. A new flotation plant was also provided. The concentrate pipeline had reached its maximum capacity, and a new line was laid. Additional concentrate storage and filtering and drying facilities were required at the port site. In Tembagapura, the 57K expansion provided for a new office complex, a staff recreation club, a commissary and more housing.

We also had to secure $716 million worth of financing. Our lead banks would see to the financing of the expansion, but they required that we first have firm contracts—or letters of intent—for the sale of at least 75 percent of our concentrate. At the time, our contracts with our Japanese buyers were expiring, and we would have to negotiate new long-term agreements. Over a four-month negotiating period we were able to place 400,000 tons of concentrate, which still left us short

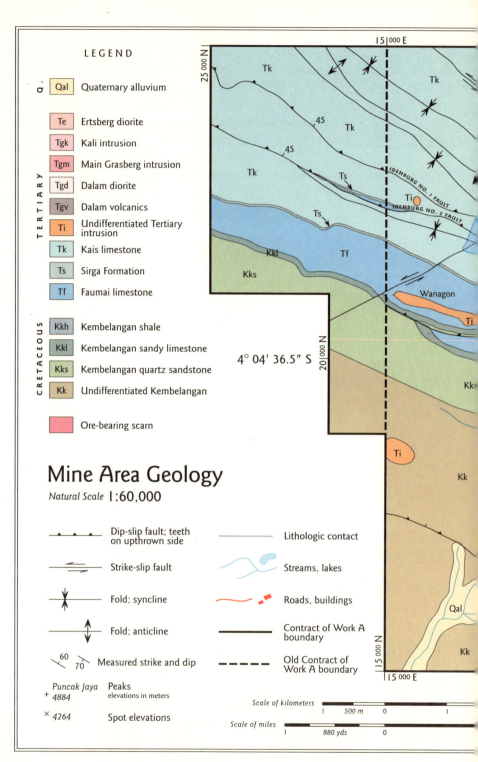

MAP 5.1

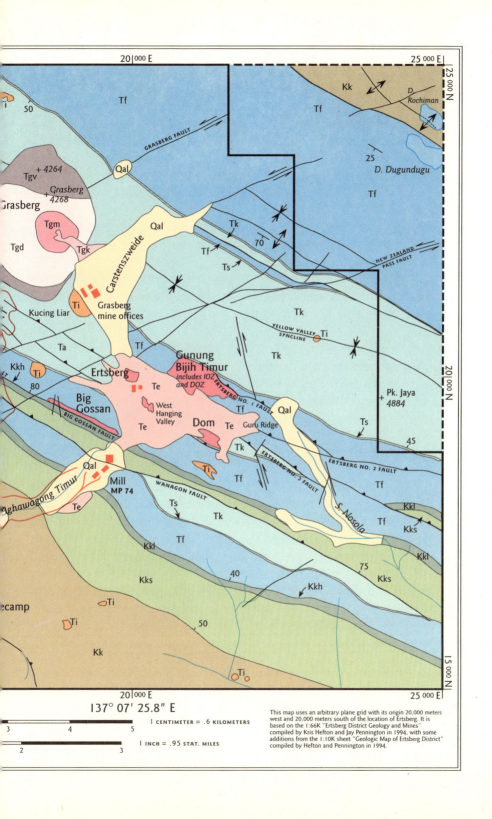

of 75 percent. Mitsubishi then approached us seeking a 100,000-ton contract to supply a new smelter they would be building in Texas City, Texas. In a relatively short period of time, we were able to agree on a letter of intent. These contracts satisfied the banks.

The banks arranged a series of presentations in the major financial centers of the world: New York, Tokyo, and London. These proved successful, and by the August 1989 annual board meeting of P.T. Freeport Indonesia and Freeport-McMoRan Copper & Gold, all elements were in place and the project, complete with financing, was approved by the board.

The only obstacle left was to successfully complete what was probably the single largest heavy construction program in all of Southeast Asia. We chose Fluor Daniel to be the design engineer and construction manager of the project. Rather than lose our 800-man construction force that was already on site, however, we developed what we call a "salt and pepper" organization. This created a construction force made up of both Freeport and Fluor Daniel personnel in which, for example, a Freeport employee might report to a Fluor Daniel supervisor.

This turned out to be an excellent arrangement. It places responsibility for the design and engineering of the facilities on the owner's representatives, in this case, Freeport. This allowed us to build our mine and mill additions according to designs that, through experience, we knew would work for our purposes and could be completed at the lowest cost. Under a normal contract, Fluor Daniel would face the usual liabilities, which could lead to over-engineered facilities, and costly and time-consuming construction methods.*

Freeport engineers were placed in Fluor Daniel offices in Redwood City to oversee design and to authorize purchasing. Once purchase orders were ready to be issued, they were forwarded to Freeport's offices in New Orleans, where the purchases were completed and the delivery to Irian Jaya organized. Freeport logistics forces would then unload matériel at the port, and deliver the construction items to the work site.

---

* It is very important in a salt-and-pepper organization to have the same owner's representatives involved from the initial feasibility studies through start-up. Otherwise, when personnel change, commitments to cost estimates and schedules are lost, as is the overall workability of the flow sheet. This system worked so well for us in 1989 that we have used it ever since.

As drill data continued to come in, we began to learn more about the geological structure of Grasberg. Like the Ertsberg diorite, the main intrusion lying under the Ertsberg, GBT and other skarns south and east of Ertsberg, the Grasberg ore body is also a three-million-year-old intrusion. But unlike the skarns, the Grasberg ore is contained in igneous rock called porphyritic quartz diorite. The stockwork ore is the part of the intrusive rock that has been enriched with copper and gold.

The word "porphyry" comes from the Greek "porphyreos" which means purple or reddish. In ancient times the term referred to the much sought-after imperial porphyry, which was used as a decorative stone. Imperial porphyry was a reddish-purple granite, laced with streaks of white feldspar. Today, the term applies mainly to rock whose texture shows a distinctive mix of coarse and fine grains, a result of its rate of cooling.

What make the Grasberg deposit unique are its scale and its very high copper and gold content. There are bigger deposits, but they have lower combined concentrations of these valuable metals. The Grasberg porphyry is one of several similar copper-gold finds in Southeast Asia, but because of its size and richness it—more than any other recent discovery—has sparked exploration in the region.

Planning a pit strategy for mining Grasberg required that our geologists map the structure of the deposit's mineralized zone. To do this, they had to develop a theory of Grasberg's formation. Drilling indicated that some of the highest-grade mineralization was located in the center of the deposit. But geological theory dictated that there should be at least some mineralization in the contact zone between the edge of the intrusion and the surrounding limestone. In other words, an intrusion this large should have created some skarning in its periphery, the way the Ertsberg diorite created Ertsberg and the GBT deposits. If substantial skarn could be found, it might require a modification of the pit shape to economically extract the ore.

As more and more drill results came in, geologist George McDonald and his staff fine-tuned their theory of the formation of Grasberg. They knew, from the varying geological "signatures" of the rocks, that the deposit was the result of not one, but three distinct intrusions of magma. Still, the lack of any indications of skarn around the perimeter of the deposit's upper surface was troubling. Finally, once deep drill results came back from the Ayam Hitam tunnel, they found skarn. But this was even more puzzling: Why, with all the limestone around Grasberg, do you have to drill down a full kilometer to find skarn?

McDonald is credited for conceptualizing what is now considered the accepted answer to this mystery. Three million years ago when the first intrusion arose, the surface topography was much lower. Grasberg didn't rise up *into* the limestone mountains, it rose *above* them—as a volcano. There is no skarn surrounding Grasberg, because in those days there was no limestone around it, only air.

Grasberg first arose above a much lower Carstenszweide some 3.5 million years ago in an intrusive phase called the Dalam Diatreme—*dalam* means "inside" in Indonesian, and "diatreme" is a type of volcano that spreads at the top. This is confirmed both by the lack of surrounding skarn, and the differential texture of the rock. Below the level of the Carstenszweide, the Dalam Diatreme has the texture of standard intrusive rock. As one moves higher, the rock acquires the texture of volcanic rock. At the very top of Grasberg, the structure of the rock suggests that it once poured from the vent of a volcano. The Dalam Diatreme was probably not formed in a single event, but built up over time, in the manner of a modern volcano such as Mount Pinatubo in the Philippines. (See FIGURE 5.1, opposite.)

The Dalam Diatreme was only the first phase of intrusion. Sometime later a second phase occurred, coming up in almost the exact center of the first. There seems to be no evidence that this event, called the Main Grasberg Intrusion or MGI, ever broke the surface. It was just a roughly cylindrical intrusion contained by the Dalam Diatreme. The MGI forms the very richest part of the deposit.

Right after its intrusion, the MGI was shattered and hydrothermally altered, perhaps by evolving aqueous fluids within the rock. This formed a network or stockwork of quartz veins that carried the highest-grade copper mineralization found in Grasberg.

A third and final phase of intrusion, called the Kali Phase, came up as a vertical plane along existing fractures, leaving a structure called the Kali Dikes. This intrusion is named after the small stream at Grasberg's base where Potter drove his first disappointing hole. The rocks left by the Kali phase are only weakly altered and mineralized, and contain mostly below-ore-grade amounts of copper and gold.

These intrusions have left an area of mineralization shaped roughly like an inverse cone, 2.3 kilometers by 1.7 kilometers near the

Formation of the Grasberg Complex

Source materials from Lawrence Johnson of the Tembagapura Block A geology office.

SEQUENCE OF INTRUSION

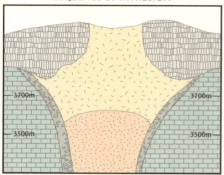

1. Dalam Diatreme. Andesite and diorite phases, and development of maar volcano.

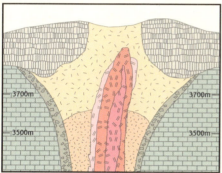

2. Main Grasberg Intrusive (MGI). Early, middle and late phases.

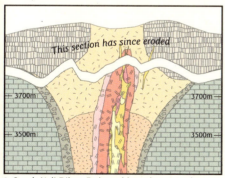

3. South Kali Dikes. Early and late phases, and erosion level.

SEQUENCE OF MINERALIZATION

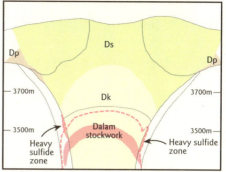

1. Dalam mineralization. Formation of deep copper-gold porphyry stockwork system.

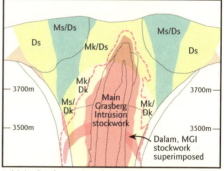

2. Main Grasberg mineralization. Development of high grade copper-gold stockwork system.

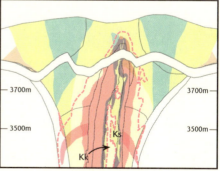

3. South Kali mineralization. Development of low grade intrusive core.

Legend:
- Late South Kali
- Early South Kali
- MGI late phase
- MGI middle phase
- MGI early phase
- Dalam volcanics
- Dalam andesite
- Dalam diorite
- Marble breccia
- Limestone
- k — Potassic alteration
- s — Phyllic alteration
- p — Propylitic alteration
- MGI stockwork
- Dalam stockwork
- D — Dalam alteration
- M — MGI alteration
- K — Kali alteration
- 1% copper limit

FIGURE 5.1

top, at 4,100 meters, and narrowing to about 900 kilometers at the 3,000-meter elevation level. From that point the intrusion narrows further to a diameter of 500 to 600 meters. There are still very worthwhile ores at 2,650 meters, the deepest drill hole to date. The copper mineralization occurs chiefly in chalcopyrite, although there is some bornite as well. The very highest copper and gold values occur between 3,550 and 3,350 meters. The correlation between the copper and gold grades is very high throughout the ore body.

Below the reach of the deepest drilling to date, there is a big question mark. Grasberg's ore is "open at depth," meaning that nobody knows how far down the deposit continues below the 2,650 meter mark. But it is unlikely that the ore body stops cold at this level. Deep drilling will be able to be conducted when the Amole adit, a six-kilometer tunnel from just above the mill, reaches the center of Grasberg in mid-1996. But there is really no rush to answer the question of the final extent of Grasberg's ore. It will take more than four decades to exhaust the already proved and probable reserves of copper and gold contained in the mountain.

More important than allowing access for deep drilling, the Amole—the name comes from a greeting in the local Amungme language—will serve as a de-watering drift, to remove any water that accumulates in the growing pit. It will also someday contain an ore conveyor system to carry Grasberg ore, and ore from other deposits, directly to the mill.

WHILE THE FANTASTIC recent growth of Freeport Indonesia could never have taken place without the discovery of the Grasberg ore reserves, there were and still are additional exciting prospects in the area. Grasberg has a tendency to dwarf everything else, but "Block A," the 24,700 acre COW around Ertsberg and Grasberg, continues to offer one of the most exciting geological settings in the world. (See MAP 5.1 pages 152–153.)

Freeport's Block A lies on the Irian–Papuan Fault Belt, at a location that has seen some unusual geological activity, especially extensive cross-faulting. Faults offer the areas of least resistance to the upwelling of magma from beneath the earth's crust, and hence are the prime areas where minerals, concentrated from these magmas by further activity, can be found. Freeport's Block A lies at the intersection of several faults, chief among them the Big Gossan Fault, the

Fairy Lakes Fault, the Grasberg Fault, the Hanging Wall Fault, the New Zealand Pass Fault, the Wanagong Fault, and the Yellow Valley Axial Fault.

Our geologists now believe that deep below the surface, this complicated pattern of faults can be reduced to just two main northwest-southeast systems: the Idenburg–Ertsberg fault system and the Wanagon–Big Gossan fault system. It is very likely that the mine area's high mineral concentration is a direct result of the crossing of these systems several kilometers beneath the earth's surface, and their intersection with the southwest-northeast Grasberg Fault and New Zealand Pass Fault.

Block A has been blessed with not one but two very, very rich Pleistocene Epoch intrusions yielding the Ertsberg and the Grasberg mineralized systems. The Ertsberg intrusion, which occurred approximately 3.25 million years ago, yielded Ertsberg as well as the GBT, IOZ and DOZ complex, and the Dom deposit. All of these are skarns, and the source of their mineralization was the Ertsberg diorite. (See TABLE 5.2, page 162.)

As Ertsberg's reserves steadily dwindled—and before Grasberg was discovered—the skarns formed by the Ertsberg diorite were critical to the company's success. The Deep Ore Zone of the GBT deposit was first investigated in 1981, with drilling to delineate the boundaries of the deposit going on until 1987, when mining began. The drills freed up from DOZ work were then directed upward from the DOZ and downward from GBT, eventually establishing the boundaries of the Intermediate Ore Zone. Taken together, the richest part of these ore bodies extends from the surface, around 4,100 meters, to at least the 2,700-meter level, at a 60–85 degree dip. Drilling has reached the 2,700-meter level, already below the level of the mill—2,800 meters—and it is highly likely that mineralization extends even further down.

In 1968 Del Flint also noted the surface showing of <u>another ore body: Dom.</u> Just one kilometer south of the GBT complex, dark outcrops of copper mineralization, stained with green, mark the rock face. Nelson headed up the initial Dom drilling in 1976. The program consisted of six holes, four of them in a fan-shaped pattern, and an additional one on either side. All of them intersected ore.

"We got good, long intersections, almost pure chalcopyrite with a lot of oxidized patches, not bornite or chalcocite such as we have over on the GBT," Nelson said. "So the grade was going to be lower than GBT, but it seemed obviously like a resource of value."

Delineation drilling from 1985 through 1989 outlined the Dom ore body. It has a wedge-shaped structure, which in cross-section looks a bit like a double-rooted tooth. It outcrops at 4,300 meters, approximately two kilometers east-southeast of the Ertsberg open pit. The Dom deposit extends nearly vertically from its surface outcrop down at least 750 meters, the deepest drilling up to the present. Helicopter-supported drilling continued until 1985, when a drill-support road was pioneered to the site and a diamond drill adit was driven. Production development on Dom was largely completed by the late 1980s, and mining had been scheduled to start in late 1991. But the discovery of Grasberg's huge ore body put a hold on the development of this resource. Dom's 31 million metric tons of ore are now held in reserve, to replace GBT when it is depleted.

Stewart and Nelson identified two additional potential deposits: Wanagong, at an alpine lake west of Grasberg, and Big Gossan, near the mill. These were mapped and surface samples were collected and analyzed. Wanagong was drilled as a gold target in 1980–1982, but the results were inconclusive.

Interest in Big Gossan picked up under Freeport's regional exploration program in 1990. The ore body is located close to the mill—but this is "close" only on the map. Although they are less than a kilometer apart, a high, nearly vertical ridge separates the mill, at the head of the Aghawagong Valley, from Big Gossan, in a valley that drains west towards the West Aghawagong.

The deposit is a skarn in the Big Gossan–Wanagong trend, lying on the Big Gossan Fault. The name comes from a geological term, "gossan," which means "iron hat" in German and refers to an oxidized rock surface covering sulfide-bearing ores. Field exploration of this deposit was made extremely difficult by the local topography.

The extent and richness of the Big Gossan deposit was established through a 10-hole drilling program begun in 1991. Because of the extensive helicopter support required, these holes were astronomically expensive, averaging $220 per meter of core, at least twice the usual cost. Though expensive, the program was worthwhile. The fourth hole came up with an ore-grade intercept averaging more than 10 percent copper, at which time another rig was brought in. The nine-month program outlined a potential six-million metric ton ore

Some $716 million in capital investment, which included bringing these large Caterpillar trucks to Irian Jaya, was required to bring Grasberg into production.

resource, lying 300–800 meters below the surface. Chalcopyrite was the dominant copper-bearing mineral, and the average assay was very high: 5 percent copper, and 2.9 grams per ton of gold.

Drilling continued on Big Gossan when the Amole Adit, leading from the mill to Grasberg, encountered an unexpected eastern extension of the deposit. The Amole crosses Big Gossan about 650 meters south of where the systematic surface drilling took place. From the tunnel—where drilling is much cheaper—results have shown a copper equivalent of 3.06 percent, and the program is continuing. Big Gossan has the highest ore grade of Freeport's stable of deposits, but its estimated 37.3 million tons are dwarfed by Grasberg.

Since 1990 exploration in Block A and Freeport's other concession areas has been actively proceeding, which is a dramatic change from the early 1980s. The new exploration, mapping and sampling program also led to re-evaluation and drilling at South Grasberg and Lembah Tembaga, where copper-mineralized porphyry was discovered in 1994.

Thick vegetation, piles of moss, mud, slippery rock, and impossi-

Proved and Probable Reserves *as of December 31, 1995*

Deposit	Metric tons of ore THOUSANDS	Average ore grade per ton			Recoverable resources		
		Copper PERCENT	Gold GRAMS	Silver GRAMS	Copper BILLION POUNDS	Gold MILLION TROY OZ.	Silver MILLION TROY OZ.
Grasberg							
Open pit	1,152,955	1.06	1.27	2.90	22.1	34.0	50.6
Underground	604,318	1.20	1.10	3.79	13.1	15.4	34.7
DOZ	49,674	1.72	0.94	8.45	1.6	1.2	7.4
Big Gossan	37,349	2.69	1.02	16.42	1.8	0.9	9.3
Dom	30,892	1.67	0.42	9.63	0.9	0.3	5.3
IOZ	20,886	1.64	0.53	9.12	0.7	0.3	3.4
GB underground	3,170	1.67	0.47	7.87	0.1	0.0	0.4
Total reserves	1,899,244	1.17	1.18	3.78	40.3	52.1	111.1

TABLE 5.2

ble slopes combine with the generally cold, wet weather to make exploration in Block A a challenge for geologists. "This is the most dangerous place I have ever worked," says geologist Kris Hefton, who did much of the recent exploration on the steep terrain surrounding the Big Gossan prospect. "I have never worked in conditions so rugged."

When we expanded our exploration program in 1990, Hefton came to Freeport from the hot, dry deserts of the western United States where he had been looking for uranium and copper.

"I felt like I had landed back in the Jurassic Period," he says. "With all the tree ferns, I expected to see a dinosaur at any moment." The rock faces in the area are so steep as to be almost surreal, and Hefton says that at first, looking down from the helicopter, his mind would adjust the incline to something more normal, even though that made the waterfalls fall at a crazy angle. But you could never get him to leave now: "What a fantastic place to work! There's such a mystique about it!"

Even though it might seem that the Block A area is so rich that one could just collar a drill hole anywhere and hit copper and gold, the geologist's job is to comb the ground and come up with specific targets. To do this means working from a helicopter, a demanding process for both geologists and pilots alike.

Once, having just touched ground after hoisting down from a hovering helicopter, the wash from the rotors knocked down a large tree branch, which bounced off the ground straight up into Hefton's face. "That sure sobered me up," he says. In many of the drop-off areas around Big Gossan, the terrain was so steep that the helicopter could only plant one skid on the ground. In these cases, the pilot has to keep a close eye on his inboard side, to make sure the blades don't hit the cliff face, and can't pay much attention to the loading and off-loading of geologists and their crews. Once, in just such a situation, Hefton was still stuck on the skid when the helicopter took off. His choice was either to ride back to camp James Bond–style, clinging to the skid, or to jump. He quickly chose the latter, and luckily landed on some relatively flat rock.

Survival skills are essential here. Sometimes the weather turns sour suddenly, and there's no way to get a helicopter back in. And it's always too far and too steep to walk. In such cases the geologists pull out their plastic tarps and space blankets, and wait it out. "As long as you stay dry, you're okay," Hefton says. "As for food, once the initial hunger pangs are over, it's no problem." But in the mountains of Irian Jaya, the geologist's most essential piece of survival "equipment" is his Irianese crew. "Don't leave home without one," Hefton jokes. He has never ceased to admire these tough men who, with just a few week's training, ride choppers and handle equipment and sample bags like professionals.

The only problem, Hefton says, is when a cuscus shows up. The cuscus is a small marsupial much sought-after for its fur, which is made into headbands and other ornamentation, and its succulent meat. These animals, which have insufficient fear of man and do not run particularly quickly, have been totally hunted out near the villages. But in the remote areas where the geologists work, an occasional cuscus is encountered. When this happens, Hefton says, "say goodbye to your Irianese crew."

Once, stuck out in an isolated area, Hefton's Irianese assistant spotted a cuscus and was gone in an instant. At an elevation of close to 4,000 meters, he ran the animal down, barely getting out of breath.

The tramways have been superseded by the efficient ore pass and conveyor systems for carrying ore, but they still haul workers from the mill to the mine site.

Then, even though it had just rained and all the wood was sopping wet, he quickly started a fire with a strip of rattan, and deftly skinned and roasted his catch. Hefton passed on his share of the meat, but was duly impressed by his partner's skill and physical toughness.

CHAPTER SIX

To 115,000 Tons a Day

The Heavy Equipment Access Trail—The 115K Program—Leveling a ridge—The SAG mill—A road through a swamp—Raising $1 billion—The privatization program—Creativity and risk—Record production levels

A STEADY RUMBLE issues from the fog long before the first bulldozer appears. Ilyas Hamid, radio in hand, watches as a caravan of heavy machines edges closer. A train of bulldozers, some pulling, another pushing, haul a steel sled up this winding dirt road. Secured to the sled is a massive steel bucket for one of the mine's power shovels. The road, as the crow flies, is five kilometers long, but in its last three or four kilometers it gains 1,300 meters in elevation, and the required switchbacks increase its on-the-ground length to ten kilometers. It takes bulldozers with a combined engine output of 2,200 horsepower to work the load up this remarkable trail, and in the course of just one trip, 100 kilograms of steel will be ground off the sled. Walking behind it, the heat from its recent passage can be felt through a pair of heavy boots. This is the Heavy Equipment Access Trail, universally known as the "HEAT road."

The shovel bucket—which has a capacity of 42 cubic meters—requires a Caterpillar D11N, two D10Ns and a D9N to drag up the HEAT road. Loads much heavier than the one Hamid now leads have made it up the hill. The turntable of the largest power shovel, the P&H 4100, is perhaps the heaviest: its weight of 93 tons, combined with the sled's 37 tons, means dragging 130 metric tons of dead weight up to the Grasberg mine. In such cases Hamid has to add another D11N—and 770 horsepower—to his powertrain.

As the machines growl toward one of the curves, Hamid pulls out his radio and tells the driver of the lead D11N to head to the right of the road, where the substructure is a bit more solid. Once the last dozer rounds the curve, Hamid jumps into his Hummer, a civilian version of the U.S. Army's rugged Humvee off-road vehicle, and drives to the next difficult stretch to shepherd the convoy through. Well over two hundred loads have made it up the HEAT road under his care, and he knows every inch by heart. This is not that surprising, however, since he built it.

Without the HEAT road, the expansion of production rate to 115,000 tons of ore a day—and even more—would not have been possible. The tramway system is limited to loads of less than 15 tons; the giant electric shovels and 218-ton trucks that make it possible to feed the mill at such a high rate cannot be broken down into such small loads. And without them, production would have become inefficient well below the 115,000 tons a day rate, and our ability to economically mine the huge Grasberg deposit would have been seriously compromised.

When those first remarkable core samples started coming back from the Grasberg drilling program, we knew that we had to expand the scale of our operations dramatically. The operation changed, almost overnight, from one that was thinking about someday shutting down, to one that was furiously planning a wide-ranging expansion that would lead to a doubling of production, and a future well into the next century. Each ore body has an optimum tonnage that can be economically mined, whether by open pit or underground methods. In the case of the Grasberg ore body, reaching this optimal tonnage required a large open pit, and after a hiatus of five to six years, when most of the ore flowing into the mill came from underground, we would be returning to an open pit operation. People would need to be trained in yet another new technology.

Throughout the underground era, from 1981 to 1988, as our proven ore reserves increased to 100 million tons, we increased production from 7,500 tons a day to 20,000 tons a day. By late 1988 it became clear that the Grasberg deposit would allow a considerable increase in production, and various mill improvements bumped the production rate to "32K," then "57K," then "66K." We knew we were facing a moving target as the reserves expanded. We did not know then—and still don't know—where it will stop. In late 1988, Grasberg looked like a 600-800 million ton deposit, and we began the 90K expansion. This project did not keep its name for long. With increased drilling and geological work, it soon became clear that there were more than 1 billion tons of ore in Grasberg. This is when the 115K was born.

Money aside, the biggest challenges of the 115K expansion were finding a way to get the equipment and materials to the mine site in the required time, and finding an area large enough at the Mile 74 mill site for building a much larger concentrator.

The equipment remaining from the Ertsberg open pit operations, decommissioned in 1989 when the ore was exhausted, was grossly inadequate for the job of mining Grasberg. The high stripping ratio at Grasberg, which averages more than three-and-a-half tons of waste rock, called "overburden," for every ton of ore delivered, means that a tremendous amount of rock needs to be moved at a very high rate if mining is to proceed profitably. Very heavy mining equipment—the largest available in the world—would have to be used. But at the time the project began, there was no way to get the big shovels and trucks to the mine site.

The challenge we faced in expanding the mill facilities was not creating access, but creating space. The grinding and flotation facilities sat at the head of a dead-end, steep-walled canyon, and there was simply no more unused ground left. A new crusher or ball mill added to one of the plants would not be enough to reach 115,000 tons a day. We would need a new plant, and one based on new technology. Feeding this plant would require a new stockpile, and doubling the mill's water supply.

The HEAT road and the mill expansion were among the most difficult aspects of the project, but challenges were everywhere. A new system of ore conveyors and ore passes would have to be built, workers would have to be hired, trained and housed, and financing would have to be secured—almost one billion dollars' worth.

In 1988 the drilling crews that were bringing back such encouraging information from the Grasberg deposit were entirely supplied by helicopter. Fuel, diamond drills, and other heavy supplies were all airlifted in, which was very expensive. Still, nobody knew for sure if it would be a working mine. Hamid was sent out with a bulldozer crew to see if he could build a road from the old Ertsberg open pit maintenance shop across the Carstenszweide to Grasberg. He had a look at the terrain, and started creating a bulldozer trail. He had only built about two or three kilometers of this first tentative trail when the news was confirmed: Grasberg was big, and had a lot of gold. It was going to be a mine.

At this time I flew out to the mine site, and together with general

Powerful bulldozers pulling a load up the heavy equipment access trail—the HEAT road—a tricky, winding path leading from the mill to the top of Grasberg.

manager Tommy Williams, mine manager Dave Francisco, head geologist Dave Potter and Hamid, took a helicopter to the top of Grasberg. We all walked down towards the area that would eventually become the entrance to the pit. Williams and Francisco asked Hamid: "Ilyas, can you build a road to the top of this mountain?" Looking down along the Grasberg slopes and the swampy Carstenszweide, Hamid said: "As long as you don't need it in a big hurry, I can do it." All he asked for was a single bulldozer, a mechanic and eight men, and he set about building the first access road to Grasberg.

Then, as samples looked better and better, and the proven ore reserves swelled, we became more and more impatient to get the road through. Hamid was doing his best, but he could only work from one end. And the smallest piece even the smallest bulldozers could be broken down into still weighed one ton, beyond the capabilities of our helicopters at that altitude. At one point our impatience was such that we toyed with the idea of renting a very high-capacity Russian helicopter just to lift a D4 bulldozer to the top. There, with gravity on its side, it could work much more efficiently pushing a road downhill.

Although spanning a distance of just five kilometers as the crow flies, the terrain is so rugged that the HEAT road covers twice this distance on the ground.

Eventually, we decided the best solution would be to set a heavy concrete anchor at the top, from which to winch a bulldozer up.

Our helicopter pilots began air-lifting sacks of concrete to the top to build the anchor for the bulldozer-mounted winch, but before they could finish, Hamid and his crew made it to the top. "When I got there," Hamid remembers, "one pilot said, 'Ilyas, you've taken my job away!' But he was just joking." The drill crews were relieved that Hamid had made it, as from then on they could get diesel fuel without having to request expensive, and unreliable, helicopter time.

Hamid's first "road" was a rough bulldozer track, but it was enough to get supplies to the drill crews. The time soon came, however, when a proper road was needed. This would need to be a minimum of 12 meters wide, and with a grade of no more than 8 percent.

"My new boss came up to me and asked, 'Ilyas, can you do that?'" Hamid said. "'I can,' I answered, 'I need a drill, a dozer with a driver, and a mechanic.' Then my boss asked, 'What would be your budget?' But I'm stupid—I can't add up the costs. We were sitting on some rocks up there on the mountain, and he said, 'There's a contractor here from America, and he is asking ten million dollars to widen the road. Another contractor from Australia is asking ten-and-a-half million. How much would you need?' 'Well,' I teased my boss, 'if the equipment is there, just give me half a million dollars.' He laughed."

Hamid and his crew widened the road, for a fraction of the cost of the outside contractors. As soon as he finished, we could bring up the "triple sevens," Caterpillar 777 trucks with a capacity of 80 tons. At that point, the large-scale stripping could begin, and we could start moving ore out of Grasberg in earnest.

Nobody at the company has been working in Irian Jaya longer than Hamid. He started out with Bechtel-Pomeroy in 1969, building the original access road. He is Minangkabau from Sumatra, an ethnic group known for the propensity of its young men to wander. Some never return home; others come back to settle down and raise a family in the house of their mother-in-law, a custom unique in Indonesia to the matriarchal Minang. Hamid, who is 56 as this is written, follows a more modern trend, and will someday settle down with his wife in Jakarta. He hopes his children, now finishing up their university education, will go on to find work with Freeport.

Hamid started off with Caltex, the oil giant, and had become an expert "cat-skinner" by the time his then-employer, Pomeroy, teamed up with Bechtel to build the access road. In 1969 there was no company jet from Jakarta to Timika, and it took Hamid several days, with an unscheduled two weeks in Singapore waiting for the airplane to be fixed in Darwin, to make it to Kokonau. From there, the workers flew by helicopter to Amamapare, and then to the Uteki River valley, which was to become Tembagapura. Their first job was to level a spot, using hoes and shovels, so the helicopters could land a disassembled D4 bulldozer for Hamid.

Most of the road-building crew found the job difficult, and of the dozen in his group, only three lasted the first year. The work was strenuous and frustrating, there was no electricity, and everybody shared one tent. Australian food—steaks, chicken and milk—was plentiful, but rice was served only once a week. Hamid accepted all this without complaint, but he admits that the lack of *cabé*, the fiery hot chili sauce essential to Minang cuisine, was hard on his morale. American Tabasco sauce for him was child's play. When the first shipment of *cabé* made it to Irian Jaya, his spirits lifted considerably.

The road from the old Ertsberg shop to Grasberg allowed work to begin, but a bottleneck remained—the tram, and its 15-ton weight limit. What we needed was an access road from the mill, or one from Ridge Camp, just downhill from the mill. This would provide the last remaining connection from port site to the Grasberg pit. One look at the path this road would have to take would be enough to frighten off any ordinary road-builder. But not Hamid. He relied on his experience and unerring instincts with a bulldozer to "feel" his way down. The work began in October 1990, and was finished in July of 1992. Hamid and his 17-man crew completed the 22-month job of building the HEAT road with only two lost-time accidents.

The HEAT road was the largest single contributor to the expansion. By October 1992, we could start bringing up the heavy gear, and the 218-ton Caterpillar 793s and the big hydraulic shovels came up in November. At first, the grade of the road exceeded 22 percent in areas, but these have all now been cut back to 15–20 percent.

Hamid personally suffered one mishap during the road-building, which cost the project four months. "I knew we had to hurry and I was doing my best," Hamid said. "But then I had an accident. I was directing a dozer and a tree fell on me. It wasn't a big tree, but it broke my arm." Hamid was sent to the hospital in Jakarta, and was out for four months.

Welders working on the bucket of a P&H shovel. The high production rate of these machines means that they must constantly be maintained.

The flood of Grasberg ore brought a rapid expansion of the milling and concentrating facilities. Here workers build forms and tie rebar for a new tailings thickener.

With the completion of the HEAT road, we began to assemble one of the largest fleets of heavy-duty mining equipment in the world. The electric shovels consist of four P&H 4100s (42-cubic-meter buckets) and two P&H 2800s (34-cubic-meter buckets), and the hydraulic shovels consist of three Hitachi EX3500s (20-cubic-meter buckets) and three Hitachi EX1800s (10-cubic-meter buckets). The truck fleet supplied by these shovels consists of 45 Caterpillar 793s (218-tons), 21 Caterpillar 785s (130 tons) and 13 Caterpillar 777s (80 tons). Ten huge Caterpillar D11N bulldozers, three D10Ns, and three D9Ns shape and clear the benches. Together with eight new drill rigs and two new crushers, the total cost of the equipment for the 115K expansion came to more than $200 million.

Moving all this ore would be meaningless without a way to efficiently process it. And reaching a sustained, daily throughput of 115,000 tons of ore would require a significant expansion of our mill facilities at Mile 74. But Mile 74 is deep in a canyon, and our existing crushing and concentrating facilities occupied literally all the available level ground. Standing in the way was the Amole Ridge, so our crews moved 7.5 million metric tons of glacial till to create a level, structurally sound footing for the new plant. And, planning for the future, the removed material was trucked to a fill location further down the canyon, where it could support additional plant components, should an increase in ore reserves justify future expansions.

The Mile 74 expansion was based around a new semiautogenous grinding (SAG) mill. This device, with a variable speed drive, provides a very flexible and economical way to mill ore. The new circuit, consisting of the SAG mill and associated ball mills and flotation cells, processes 55,000 tons of ore a day. It has been running since February 1995, and its performance has been excellent. By the middle of the year, our SAG circuit was processing more ore per day than any other SAG mill in the world.

"It's not that it has run like a Swiss watch," observes mill manager Charlie Wilmot. "SAG plants don't do that. But it has run very, very well and has met all of our expectations. And we've had virtually no nasty surprises with equipment breakdowns or the operators causing us major problems."

One reason the startup was so timely and trouble-free was that Wilmot brought in the 200 new operations, maintenance and technical staff workers six to twelve months early. This way the men could go through a full training program before the new equipment was installed. In staffing the SAG mill, Wilmot chose a mix of experienced and new personnel, making for a smooth transition.

The HEAT road made it possible to get the heavy equipment to Grasberg, and the new SAG plant gave the mill the capacity to process the ore these machines could move. But we still needed a high-capacity ore conveyor system to get the ore from the mine to the mill. The horizontal distance between the Grasberg pit and the mill is only 4.25 kilometers, but the terrain drops almost a kilometer in this short distance—from 3,885 meters at the first ore crusher to 2,900 meters at the mill stockpiles.

The existing ore passes and conveyors, backed up by the cable tramway, did not have the capacity to transport the additional ore required by the 115K program. The ore tramways in particular were becoming more of a problem than a solution, and we needed a completely new system for Grasberg.

The Grasberg ore flow system, which cost close to $250 million, is a marvel of innovative engineering. From two crushers beside the

Grasberg pit, the rock drops through two downhill conveyors to the level of the Carstenszweide. There dual 84-inch belts carry the ore to a set of ore passes, where it drops 550 meters into solid rock ore bins, and from there is carried on another system of dual belts, finally to be discharged onto the mill stockpiles. The first four ore passes, two meters in diameter, began operating in 1988 and cost $5 million each to construct. Three more, which exceed three meters in diameter, have since been added. With the new system, the ore trams were no longer needed.

The ore flow system has sufficient redundancy built in to allow parts of it to be shut down for regular maintenance, and even extended repairs, without interrupting production. Each of the parallel belts in the conveyor network can maintain production at mill capacity if needed, thus allowing the other to be shut down. Of the six parallel ore passes, only two are needed to maintain 115,000 tons a day of ore flow.

EVERYTHING NEEDED AT the mine, the mill and Tembagapura comes up the single road linking the coast with the interior. This is the lifeline of the operation. With daily maintenance, the original Bechtel-Pomeroy-built road had served the company well, but with the expansion to 115K, and all the new equipment that would have to make the journey, the road would have to be improved. It would have to be straightened and widened, guard cables would have to be installed on the more dangerous corners, and most importantly, it would have to be extended all the way to the port.

The nine-mile stretch of road from the port at Amamapare to where the road turns west to Pad 11 had been attempted in the late 1960s by Bechtel-Pomeroy, but they were unable to conquer the coastal swamp. Until the new road was built, every nut, bolt, egg and frozen steak unloaded at Amamapare had to be transferred to barges and shipped 11 miles up the winding Jaramaya River to Pad 11. There, the containers were transferred to trucks, and driven to Timika or Tembagapura. We knew from the beginning that the increase in matériel required by the expansion would overload the barging capacity of the shallow Jaramaya River.

There wasn't too much pre-planning for this road. Time was short, and the road and dock had to be complete before the large volume of equipment and supplies for the 115K expansion arrived. So

A scarecrow made from a raincoat, helmet and dust mask urges passersby to go slowly around the construction crews improving the roads in the lowlands.

we drove it south, parallel and immediately adjacent to the route followed by the concentrate pipeline, an area we knew well. It was clear from the start that it would be too expensive to build a bridge from the end of the road to the port site, so a dock was constructed where the road would end at the river's edge. Construction on the dock began at the same time as road construction, using equipment mounted on barges.

Building this nine miles or so of road cost less than $10 million, a bargain considering that it runs through a swamp. And it only has about a meter and a half of grading material in it, in sharp contrast to the road Bechtel-Pomeroy built from Pad 11 to Timika, which in places is supported by 10 meters of fill. Between 1969 and 1992 a new material was invented, called "geofabric," which acts as a barrier between the roadbed and the underlying swamp. The geofabric has proved much superior to the old method of fill and corduroy, or tree trunks. In the softest sections, the geofabric was supplemented by 30,000 worn out tires.

While the port road was being extended through the swamp, senior manager Horst Lossman was overseeing scattered improvements in the rest of the road. Between 1992 and 1994 he straightened its course as

The barracks at Ridgecamp—Mile 71—just downhill from the mill, spare workers the lengthy commute from Tembagapura or the lowlands.

much as possible, smoothing out the sharpest turns from a minimum radius of 6 meters to a minimum of 30 meters. He also widened it where possible, from 8 meters to 10–12 meters. The most dangerous stretches he lined with used tires set in concrete and linked with old tramway cable. These guardrails have proven effective. A bus carrying 64 passengers ran off the road above a 300-meter precipice—but the cables held and nobody was hurt.

Because of the high cost and because in places the diesel fuel and slurry pipelines—which require regular inspection—run underneath it, the road cannot be hard-surfaced. Lossman did pour concrete on one spot, however, near Mile 57, which until then had been averaging one breakdown a week. Because of the steepness of the terrain, the maximum grade of 28 percent could not be altered.

Heavy use and constant rain—an almost unbelievable 10 meters a year at Mile 50—mean the road must be constantly maintained. An average of 25 centimeters of gravel is laid down every year, which is equivalent to resurfacing the road every nine months. A crew of 180 men, with a yearly budget of $5 million, keeps the road in shape.

The road is the lifeline of the property, and requires daily attention. If it is left unattended for even a few days, it might take a month to get it back in shape. Landslides cause the most traffic delays. These are worst along the nine-kilometer stretch between Tembagapura and the mill, where the road rises more than 750 meters while hugging precipitous slopes. But with crews, bulldozers and graders ready day and night, even a major landslide stops traffic for only a few hours.

The road is very heavily used. In 1995, a total of 78 trucks with a capacity in excess of 20 tons, 165 trucks under 20 tons, 73 buses, and some 862 light trucks were using the road. The big trucks shuttle continuously between the port and supply depots inland. The 73 buses make 750,000 man-trips a month between the barracks and the mine. Freeport has one of the world's largest fleets of Caterpillar equipment, and in 1994 we were the heavy equipment manufacturer's single largest customer.

All of our vehicles run on diesel or, in the case of the largest shovels at the mine, on diesel-generated electricity. The electric plants generate 103 megawatts of power: 74 megawatts for the mill, 16 megawatts for the mine, 6 megawatts for Tembagapura, and the rest for the port and the other scattered installations. Keeping the equipment and power plants running takes 12,000 gallons of diesel an hour.

Except for some locally grown vegetables, nearly all of Freeport's supplies come in by ship to Amamapare. Just feeding the 17,000 people who work for Freeport or its on-site contractors for one month requires a staggering 121 tons of beef, 143 tons of chicken, 220 tons of rice, 24 tons of potatoes, 25 tons of fish and shrimp, 6 tons of lamb, and 4 tons of pork, along with 21,000 gallons of milk and over a million eggs.

The combined tonnage of supplies coming in and concentrate going out topped a million metric tons for the first six months of 1995, pushing Amamapare into the top five of Indonesia's ports. Only Jakarta and Surabaya surpass Amamapare in tonnage on a regular basis. During these six months, 684,000 metric tons of concentrate were loaded on 51 bulk carrier ships, and 132,000 tons of diesel fuel and 183,000 tons of general cargo were off-loaded from 61 cargo ships. Except for the fuel and occasional oversized loads, the incoming freight arrives in standard 20-foot containers, about 1,500 a month. Some 8,000 of these containers move around in Freeport's supply system at any one time.

The geography of the port site is far from ideal for shipping. Amamapare is built on a soggy, tidal swamp, which has been raised by fill. Worse, the shipping channel from the Arafura Sea to the port offers only 6 meters' draft, which limits the choice of vessels and controls the loading of vessels passing through the channel. In the silt-laden rivers of Irian's south coast, dredging is futile, and any man-made channel closes up faster than it can be cleared. This means that the ore ships can only be partially loaded at the dock. Then they must go back out to sea, 12 kilometers from the dock, and top off their loads from barges. About 40 percent of the concentrate shipped has to be barged to the concentrate vessels, and transferred to their holds with the ships' clamshell-like grabs. This procedure requires that it not be raining, and that the waves be no higher than two meters. On a year-round basis, two days out of three offer adequate conditions for ore loading at sea.

THE TOTAL COST of the 115K expansion, including $300 million for the mill expansion, more than $200 million for the heavy equipment, and $250 million for the ore flow system, came to $952 million. The company raised this money through the combination of a special stock sale and an extensive program of privatization, selling its many assets in Irian Jaya that are not directly production-related.

Through Freeport-McMoRan Copper & Gold, we issued commodity-linked preferred stock; that is, stock whose value is directly linked to the price of gold, in one case, and silver, in the other. The value of this stock rises and falls with the price of the commodity. Freeport-McMoRan Copper & Gold then loaned this money, borrowed from the public, to P.T. Freeport Indonesia, the operating company in Irian Jaya. The terms of this loan are the same as the commodity-linked stocks, making this a "mirror image" security. All told, the company raised about $500 million this way, funding approximately half of the 115K expansion.

A company that needs money can get it on the debt market, but interest rates can undermine the profitability of any new production facilities the loans make possible. It can also issue more stock, but this dilutes ownership, and equity financing can be even more expensive than debt financing. Our commodity-linked stock offering avoids some of these problems, but an even better solution has been the privatization program.

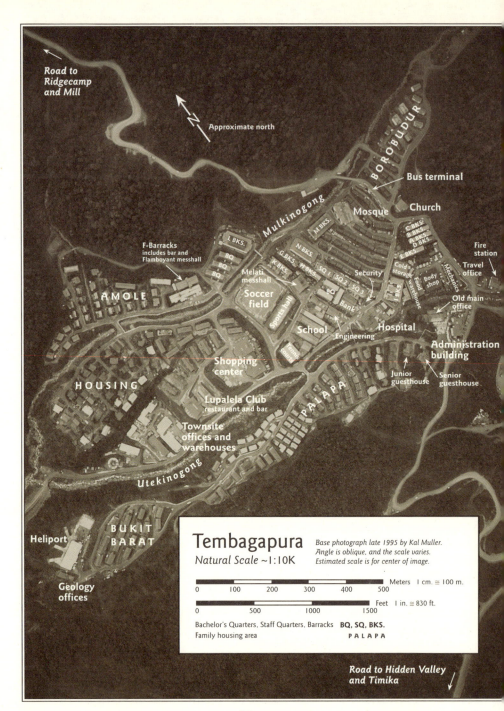

MAP 6.1

Hidden Valley, in the foreground, was built to provide apartment-style housing for workers. Dormitory space at Tembagapura, just over the ridge top, was also increased.

The impetus behind this strategy came from Freeport chief executive Jim Bob Moffett. He looked at our facilities and realized that there were other people who could run some of our non-mining operations as well—or even better—than we do. Because Irian Jaya was so remote and undeveloped, in the course of developing our mines we had to build and operate a large number of ancillary businesses, including transportation services, hospitals, power plants, hotels, housing, stores, shipping services, bakeries, restaurants and mess halls. Although necessary to the mining operations, we realized that these valuable assets could be sold, thus raising much-needed money, and operated just as efficiently by someone else. Maybe even *more* efficiently. Although the depth and range of talent at Freeport Indonesia is considerable, why should a mining company have any particular expertise in, say, running a hospital?

Another factor is risk. Freeport-McMoRan Copper & Gold was getting up to a 40 percent return on its assets. This is the kind of return a high-risk business like mining requires. A hospital doesn't return 40 percent; a mine production expansion can. We saw an opportunity to recover our capital from non-productive assets, and to put it into an investment—the 115K expansion—that would reap a high return for our stockholders.

In 1991, Kevin Pollard, an executive vice director and our manager of outside investments, was put in charge of evaluating the company's non-core businesses in Irian Jaya, and finding buyers.

His job began with selling all the buildings, the hospital, the guest houses, the family houses, the barracks, the shops, and the buildings themselves. Then he located a company to operate the catering, housekeeping, medical and other services. As Pollard worked down his list, the value grew from $20 million to $270 million. All of these were sold to a joint venture company 75 percent owned by the Alatief Corporation and 25 percent owned by P.T. Freeport Indonesia.

When we first looked at the power plant, we thought it was too close to our core operation, and were reluctant to include it in our privatization. Then, upon further examination, and after the Indonesian government began offering opportunities for foreign investment in electrical power, this possibility began to look more attractive. We were approached first by Caterpillar, whose interest was either as an operator or as a lender, but always as a supplier of equipment. Our search widened and finally, from a slate of several candidates, we selected Duke as the prime operator-investor, supported by other foreign and Indonesian investors. In total, $215 million was raised from the sale of existing power plants and other facilities already under construction. This capital was recycled to mine production facilities, and we now buy our power by the kilowatt hour like most other mining companies in the world.

Then an Indonesian subsidiary of the Singapore-based Asia Pacific Services was found to operate our medical services, and operators were found for the aviation services, the port site, transportation and logistics, the sewer systems, and the drinking water supply. In all, the total raised by privatization came to $830 million.

Although operating the Freeport facilities represents a good business opportunity for the third parties being brought in, we have been careful to insure that the services provided are up to our standards of quality and are cost-effective. We do not negotiate a deal that's bad for either party. But we want the operator to have equity at risk, and we structure the terms of our relationship so that they have the incentive to do things that are good for Freeport, and the disincentive to do things that we don't want.

It takes more than money to carry off a project of the scale of the 115K expansion. It takes hard-working, creative men who don't stop at the end of a 40-hour week. The uniqueness of Freeport's Irian Jaya operations and the company's driven style has always attracted smart, problem-solving workers.

"I've never seen a place as dynamic as this, never a place with so many talented people as here," says one newcomer. "There are no problems, only challenges. Many of the decisions have to be made under pressure, on the spur of the moment. You don't write letters and ask for approval."

It was this kind of quick, seat-of-the-pants decision-making that saved the company money and kept the expansion on track. When one of the huge electric shovels arrived, the only truck designed for such a heavy load was out of service at the mechanic's shop. To wait for this truck to be available would have meant keeping an expensive, high-capacity machine idle. The solution? Use a smaller truck. The small truck incurred $30,000 worth of damage hauling the electric shovel, but the increased production and profits from bringing the shovel up on schedule were worth millions of dollars.

Operations general manager John Macken has worked on every one of our expansions, from the first 9.5K to the 115K. All of these were handled internally, although much of the work associated with the 52K and 115K expansions was accomplished in partnership with Fluor Daniel. This contractor, with which Freeport has had a long relationship, was driven by our management as aggressively as the Freeport staff. "Risk is what you pay these guys for and they charge you for it," Macken says. "So if you accept that risk, you can really manage it to your own advantage, which we did."

The uninterrupted series of expansions leading to the 115K have kept the mine's management team lean and efficient. To drive these expansions, the company has built up a construction division that would be the envy of any large specialist construction company: 5,500 men and a $65 million fleet of equipment, including 32 cranes, some with 200-ton and 150-ton capacities.

"Our construction capabilities are equal to our mining," Macken says. "We've implemented projects here that most of the big players like the Fluors and the Bechtels are envious of. We do know our baili-

Grasberg ore pours into the stockpile from the new ore pass and conveyor system, an engineering achievement crucial to the profitability of the operation.

wick here very well; we know how to work our Indonesian crews."

Our expansion team measures its success by the calendar. Their push is to construct and bring facilities on line as early as possible. The big success story of the 115K was not so much that it was within budget, but that it was finished eight months ahead of schedule. Those eight months were worth $200 million in additional revenues, and maybe $40-$50 million in extra profit. The year 1995 was a very good one for the company, largely because the expansion was completed so quickly.

When we reached 115,000 metric tons a day, our Irian Jaya facility went from a mine that was interesting to the industry because of its remoteness and the technical challenges it faced, to a world-class mine by any measure. Right now we produce more than five percent of the world's newly mined copper.

The 115K is a whole new world, and it took a lot of work and dedicated people to get there. We went from an underground operation using 5-cubic-meter loaders and 20-ton trucks to an open pit operation with 42-cubic-meter electric shovels and 218-ton trucks. The increase in efficiency is staggering. When Grasberg mining first began, we were moving 60,000 tons of material a day with 300 men. Now, with 1,333 men, we move half a million tons a day.

The heavy equipment working Grasberg is the best in the world, and so are our miners. Indonesia does not have a long mining tradition, and we have had to recruit and train almost everyone from scratch, but the men are dedicated and capable.

The 115K expansion added workers to the payroll, but the additions did not keep up with the tonnage. Expansion in the long run doesn't bring as dramatic an increase in the labor pool as one might suspect. It takes a core number of people to run any piece of equipment, and doubling the size of the equipment doesn't double the number of people it takes to run it. At the mine and the mill, only about a ten percent increase in our work force was needed to double the level of production.

As of August 1995, the total number of workers associated with our Irian Jaya mines came to 17,308. Of these, 5,019 work directly for Freeport—1,004 staff members, including 220 expatriates, and 4,015 hourly workers, including 84 expatriate Filipino miners. The core operations, the mine and the mill, employ 2,385 and 706 people respectively.

THE EXPANSION TO 115,000 tons of ore a day has been an unqualified success. In 1996, P.T. Freeport Indonesia will produce 1.1 billion pounds of copper, and about 1.5 million ounces of gold. And it will do so at a very low price per pound of copper. In fact, according to the London-based metals market research firm Brooke Hunt, our mine is the single lowest cost producer of copper in the world. In 1994, with the expansion only partially in place, our cost to extract each pound of copper, after gold and silver credits, averaged 40.1 cents. In 1995, with the 115K complete, this dropped to 23.7 cents.

The average *daily* revenue being generated by our Irian Jaya mines is more than $5 million. And the expansion was undertaken with the future in mind. The mill additions left enough level ground for another plant, and the conveyor system and the rest of the infrastructure improvement was designed with excess capacity built in.

Today, with the expansion fully in place, our production rate has increased by a factor of sixteen since mining began in December 1972. Our concentrate production now averages 5,500 tons a day, with some days over 6,000 tons. Soon, without too much additional fine tuning, we will be producing as much concentrate as we were moving raw ore when we began to mine Ertsberg. This, more than any other statistic, offers an idea of the increase in the scale of our operations that has come about through the series of expansions that ended with the 115K.

PART III
The Business of Mining

Ontel Rumaropen at the controls of a P&H 4100 shovel in the Grasberg pit. With a 42-cubic-meter bucket, this is the highest-capacity electric shovel made.

CHAPTER SEVEN

Mining the Ore

Shrouded in fog—Four billion tons of rock—Poker chips—Running the 'Pay Hah' 4100—The crusher—Underground mill sweetener—Block cave mining—The 'rocks in the box'—Concentrators—The slurry line

THE GRASBERG MINE perches 4,200 meters up in the remote central massif of the island of New Guinea, one of the most forbidding places in the world. The air at this altitude is thin and cold, and the mine itself is almost always shrouded in fog. The men stop operating the open pit mine equipment when visibility drops below 40 meters; that is, when an electric shovel the size of a large house is not visible at 40 meters. This is a daily occurrence. Usually, in the early morning, sunshine pours down on Grasberg, revealing a breathtaking view of snow fields, jagged mountaintops and deep valleys. By afternoon, it rains and the fog comes in.

Although the miners appreciate the beauty that surrounds the mine, they do not pause to stare. There's a job to get on with: moving four billion tons of rock.

The opening of ore production from the Grasberg pit was remarkably rapid. Construction began as soon as the first drill holes started to prove out in the summer of 1988. The first task was to provide surface access to supply the drilling program and later the pioneering of the first open pit benches. The first Grasberg ore reached the mill in December 1989, just 22 months after the discovery of Grasberg. This is probably some sort of world record for the development of a newly discovered ore body. Typically in the industry, ore bodies languish five or ten years or more before development. Ertsberg, for example, waited 36 years.

From the completion in 1993 of the HEAT road, which allowed the huge shovels and trucks access to the mine, to early 1995, Grasberg ore production doubled. With this expansion came great gains in efficiency. In the beginning, Grasberg operations moved an average of 10 tons a day for each employee; today this figure has reached 150 tons a day per employee.

"We can only use superlatives," says Fred Mason, our vice-president in charge of operations, who oversaw this rapid expansion and increase in efficiency. "Our planned expansion had a storybook ending." Mason credits hard work and overtime put in by the engineers, operators and mine force for this success.

THE SERVICE TRAM from the mill drops workers off at the southern end of the Carstenszweide. The *weide* is at an elevation of 3,600 meters; the top of Grasberg reaches 4,209 meters. Once a marshy alpine meadow full of sinkholes and, in places, chest-high sawgrass, Carstenszweide is now the hub of support activity for the Grasberg mine, housing the operations center, some of the geology offices and a core shed, materials warehouses and the explosives magazines.

More than 1,000 men work at Grasberg, and the mine operations run 24 hours a day, in two 12-hour shifts, 365 days a year. To meet the production goal for 1995, many employees volunteered to work overtime. It took more than three million man-hours to get the job done, but at the end of the year they had moved 191 million metric tons of combined ore and waste. This effort yielded 1 billion pounds of copper, and 1.4 million troy ounces of gold.

The current cut-off grade for Grasberg ore is .8 percent "copper equivalent." This means ore in which the actual copper content, together with the value of the gold content expressed as a copper percentage, is at least .8 percent. Grasberg rock that meets the .8 percent yardstick is sent on to the crushers and the mill for refining. Rock that does not is placed into overburden stockpiles. The stripping ratio at Grasberg is quite high for a copper mine, and 3.67 tons of overburden must be moved for every ton of ore.

The top of Grasberg originally reached an elevation of 4,209 meters, but by the time mining is finished it will be an open pit 2.5 kilometers across north to south, and 2 kilometers across east to west, with the bottom of the pit at an elevation of 3,005 meters. At the current rate of production, it will take forty-five years to mine the proven and probable reserves at Grasberg, and since the mine is still "open at depth," there is a strong possibility of more reserves lying beneath the current drilling limits.

The Grasberg deposit is not a continuous, homogeneous body of ore. The ore deposit formed over the course of three separate sets of intrusions, which leaves it with a horseshoe shape, varying mineral content, and in some sections, has left weak rock that requires special attention when designing the pit walls.

The current designed slope of the pit walls is about 42 degrees, but the efficiency of the operation would be improved—that is, the pit would uncover more ore and extract less overburden—if the walls can be increased to an angle closer to 45 degrees. Then the pit can be driven deeper.

This is difficult because sections of the Grasberg deposit are afflicted with what has become known as "poker chip" rock—weak, fractured diorite that got its name because it falls out of the diamond drill core barrels in round flakes like poker chips. The second of the three intrusions that formed Grasberg came up quite suddenly, in geological terms, creating a series of fine fractures in the brittle rock of the original intrusion. As part of the chemical alteration associated with this second wave of mineralization, these fractures were subsequently grouted, or "healed," by anhydrite, a calcium sulfate mineral. In areas where rain and ground water have later softened and leached out the anhydrite, Grasberg's mine engineers are left with this weak, unstable "poker chip" material.

Although this is currently a problem, it will not be one permanently, as the pit will eventually extend beyond the area of poker chip rock into the stable marble and limestones that surround the Grasberg intrusion, allowing a stable pit of 45 degrees. In the meantime, the strength and steepness of the slope will be maintained by careful shaping and by insuring that benches constructed of the fractured material are kept well-drained. Draining the pit walls makes them more stable because water lubricates the natural joints, fractures and faults in the rock.

In an area with as much rainfall as Irian's highlands, draining an open pit becomes a serious matter. For example, the pit left by the Ertsberg mine is now Lake Wilson, named for Forbes Wilson. Lake Wilson serves as a useful reservoir for the operation of the mill, and approximately 2.5 megawatts a day of clean electrical power is generated from this water. It would be exceedingly unhelpful if the Grasberg pit were to begin to fill prematurely and work is well underway to ensure that it is well-drained.

By the end of 1996 the Amole Adit, which begins through a portal just north of the mill, will reach the center of Grasberg. The Amole will provide a system to drain the pit and will provide a new ore-conveyor system to take Grasberg ore directly to the mill. This six-kilometer tunnel, which cuts through some of the most interesting ore prospects in the world, will provide a platform for additional deep drilling. Nearer to the mill, the Amole passed under Big Gossan, and drilling this deposit from underground added millions of tons of ore to our reserves.

Blasting determines the basic but ever-changing sculpture of the pit. Dynamite, which is touchy, expensive, and deteriorates over time, has not been used in mining operations for almost 25 years. Today we use a powerful mixture of fertilizer and diesel fuel called ANFO—an acronym for ammonium nitrate and fuel oil. ANFO is safe to handle. You can jar it, or even drive over it and nothing happens. It takes a special detonator to set it off. At Grasberg, blasting takes place twice a week, using around 230,000 kilograms of ANFO each time. When mining began at Grasberg, the blasting pattern was set up to create fairly low benches, but with the arrival of the large machines on the HEAT road, bench height has gradually been raised to 15 meters and will eventually be raised to 20 meters.

Grasberg's ore is handled by the largest and most technically advanced mining equipment in the world. At the time of this writing, 16 loading units, 22 bulldozers, 12 graders and 80 haul trucks are working the pit. The shovels that excavate the blasted rock from the walls of the pit are the largest ever built—a bucketful on the very largest contains 42 cubic meters of rock. That's enough rock to fill an average bedroom, in one scoop. Shovels this big use electricity for power, and at peak load, digging into hard rock, each requires 3.2 megawatts of power. These are made by P&H in Milwaukee, Wisconsin, and sell for $7 million each. To deliver such a machine to the top of Irian Jaya adds considerably to its cost, especially considering that, even dismantled, these shovels have components weighing as much as 79.6 tons.

The Grasberg mine keeps four of these units busy, 24 hours a day, plus a dozen smaller ones. When the rock has been loaded, the big trucks take over. These are among the largest production trucks in the world, each with a capacity of 218 tons. The Grasberg operations keep 45 of these $1.6 million vehicles continuously on the move,

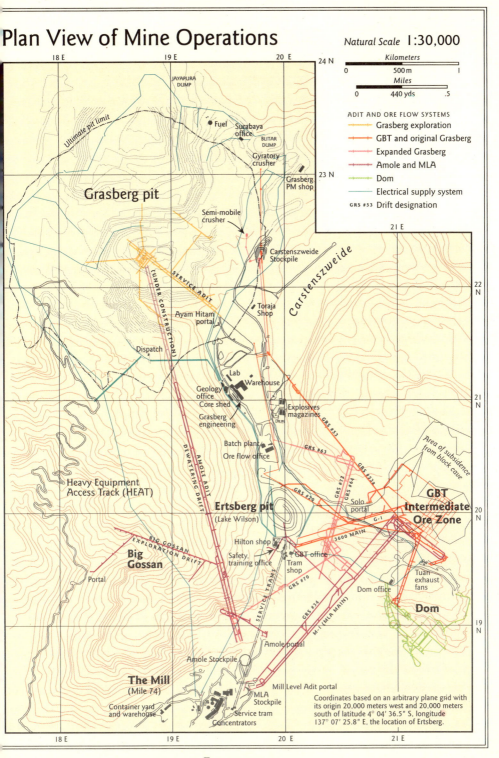

FIGURE 7.1

together with another three dozen smaller trucks. The biggest standard bulldozers manufactured also work here, Caterpillar D11Ns, ten of them at $1 million each, along with 11 smaller dozers.

The work of moving the broken rocks starts with power shovels. The P&H 4100s are essential to the mining efficiency of the pit. On a good day, positioned properly, we can get 80,000 tons from just one shovel. This is a very high-production unit. In fact, we can almost feed one crusher with a single shovel.

The operator's cab on a P&H 4100 sits high above the ground, which is where a typical shovel operator such as Hasan Naziarta will find himself. The men at the mine work 12-hour shifts, in a pattern of two days on, two days off, three days on, three days off. Access to the mine is such a time-consuming affair—a bus ride from the dormitories, then a ride up the aerial tramway, then another bus ride to the mine—that the miners prefer the longer shifts to the old eight-hour shifts. A normal work target for Naziarta might be to fill 240 trucks on a 12-hour shift, but he has exceeded this on occasion by as much as 50 percent. With Freeport's bonus system, this helps boost his salary to an average of $500 a month. This is very high by Indonesian standards, and much more than the 33-year-old Naziarta could earn in his native Belitung, an island in the Java Sea that is part of South Sumatra province.

He sits in a high-back seat with a small computer screen to his right displaying the location of the trucks. Each hand rests lightly on a joystick, like those used in video games. It takes a delicate and very experienced touch to move this huge machine with the necessary precision; the controls cannot be forced or jerked. The left joy stick moves the whole unit forward and back, opens the bottom of the shovel, and sets off a horn which tells the truck driver that his load is ready. The right stick swings, raises, and lowers the shovel. This equipment requires very good hand-eye coordination, and we have found our Indonesian workers to be very skilled in this regard.

Three buckets is all it takes to fill one of the big Caterpillar 793s. There are plenty of waiting trucks, so there's no rest between loads, no time for a quick cigarette. Fog and clouds swirl around the shovel, which Naziarta calls a "Pay Hah," the Indonesian pronunciation of "PH." Before being allowed to operate the 4100, he underwent more than 1,500 hours of training, starting with the smallest haul trucks, then the bigger trucks, then the various bulldozers, and finally the smaller power shovels. Naziarta is very conscious of safety, and he remembers one of the trainees on the 4100, who did not look before swinging the bucket and hit one of the big Caterpillars. The force of the impact knocked the huge truck sideways three meters—"What if that had been a person?" he says.

Operating conditions at Grasberg are probably the toughest at any mine in the world. Steve Drake, the superintendent in charge, notes that the daily rain and fog present constant safety challenges and add greatly to the expense of routine maintenance. For example, maintaining the service roads around Grasberg costs two to three times what it would elsewhere. The entire mine never shuts down, but dense fog cuts visibility enough to stop work in sections of the mine for 15 minutes or so at a time on a daily basis, which also adds to running costs.

Ontel Rumaropen, from Biak, an island off the north coast of Irian Jaya, is a typical driver of our Caterpillar 793s. He lives in a company-provided dormitory room with three other employees, all from Biak. By far the largest percentage of the Irianese working for Freeport come from Biak, which is a small island with few resources but plenty of people. Also, because the island was safe and easy to reach, Dutch missionaries got an early start there, educating as well as evangelizing. Biakans have been schooled now for several generations, which gives them an advantage on employment exams over most Irianese, for whom schools are very recent. Having relatives and friends working for Freeport also helps Biakans looking for a job. They've already heard about the high salaries and work conditions.

For Rumaropen, this is a comfortable job. It's not too hot or cold in the truck. With bonuses, he can earn $500 a month—that's a lot of money in Biak. The only real problem he had when he moved here was getting used to the cold. Biak is always hot and humid.

As Rumaropen pulls his truck next to the shovel, his computer screen flashes "loading." The truck shakes slightly as the shovel dumps in a load, and in minutes his truck is full. The screen, controlled by the dispatch office, tells him where to take the rock, with a message such as "TO 387 OBO," an overburden dumping location. He acknowledges the message, and drives to the site at the edge of the pit. Clouds sweep across the open area at the edge of the dump, and fog closes in as he waits his turn, dumps the rock, and heads back for another load.

One of the super-heavy-duty Caterpillar 793Bs working the Grasberg pit. These trucks are the world's largest, and have a load rating of 218 tons.

The trucks carrying mill-grade ore, from all over the pit, converge on the same machines: the gyratory crushers. This awe-inspiring device can crush up to 8,000 tons an hour of literally anything. Boulders go in; nothing larger than 200 millimeters in diameter comes out. Gary King, our manager of mine resources, says the workers call this machine Ibu Besar, which means "Big Mother." King is in charge of all the equipment and infrastructure needed to move the ore to the mill. They call the gyratory crusher Ibu Besar not only because of its size, but because it is the "mother" of the ore-flow system—this is where it all begins.

The crushers are at the 3,800-meter elevation and from there the ore travels down inclined conveyor belts, and is dropped from above onto a stockpile, which serves as a "surge pile." The conveyor belts, 84 inches wide, are the blood flow of the mine. The drive mechanism for these belts consumes three-quarters of the total 15 megawatts an hour used by the mine operations. In level areas, the belts are made of nylon-reinforced rubber, and in the downhill areas they are further strengthened with steel cord, like the carcass of a truck tire.

Reclaim feeders take the ore from the bottom of the surge pile and transfer it, through a series of belts, 3,600 meters long, to the top of the ore passes. Fine-tuning the system has resulted in an increase in the speed of the conveyor system, which now moves at 700 feet per minute. Ore flow is the key to the operation. If there is an interruption of more than a few hours, the stockpiles in front of the concentrators start to dwindle. If they ever disappear entirely, the mill stops—and so does the cash flow that supports the enterprise.

Although the ore loses some altitude as it works its way along the conveyor system, Grasberg is still so much higher than the mill that it must drop another half-kilometer through the ore passes. The amount of energy released by this free-falling rock is tremendous, and standing anywhere near the top of an ore pass, the roar and rush of air communicates a distinct sense of Armageddon. Seven of these are now in place.

The force of the falling rock turns the ore pass into a huge compressor, which must be managed carefully. The dust produced at the bottom must be contained, and if the ore flow at the top ever stops suddenly, the compressed air inside the ore pass re-expands, rushing upward. An abrupt stoppage with the system operating at full capacity can result in a blast of air that lasts 40 seconds. And this air belches forth with such energy that it brings rocks with it.

The bins at the bottom of the ore passes are designed to function best at a level of 50 percent capacity, and when all goes well, a variable-speed apron feeder, remotely monitored by video cameras and controlled by a computer system, insures that bin load hovers around the optimal 50 percent level. But the surge capacity of the bins is limited, and costly and aggravating blockages do occur. The mine crews' Christmas present in 1993 was a blockage that took 15 days and hundreds of kilograms of explosives to clear. Sometimes even technology and overtime can't provide the solution. Leon Thomas, underground mine superintendent, tells the story of how another blockage was eventually cleared:

"It was in February 1990, and ore pass number one was plugged. We blasted four or five times and even tried a diamond drill from the top. We tried everything. I was even willing to have a séance with one of the Filipinos. He told me that the problem was Santos, who had been a fatality underground before the ore pass problem. We had to clear things with his ghost, who was upset with Freeport. The crew couldn't find any black chickens, so they had to use black ducks. The guys killed them and sprayed the blood all over the ore pass. We blasted one more time and everything came down."

From the bins, a mill-level system of conveyor belts carries the ore to the stockpile at the mill. Here there are three parallel sets of belts, in two systems: the M-System—"M" for "mill"—and the Super-System. The M-System has two feeders and eight conveyors, and runs a total of 2,450 meters of belts. The Super-System also has two feeders, but its seven conveyors have wider belts and run a total of 2,300 meters of belts.

With the mill now requiring 125,000 tons every day, we can ill afford long stoppages. Our only insurance policy currently lies in an up to five-day reserve pile outside the mill, which does not give us much peace of mind.

WHILE MOST OF Freeport's attention remains focused on the open pit work at Grasberg, underground mining continues on the GBT deposits. The ore from the IOZ and DOZ is bornite-dominant, unlike Grasberg ore which is mostly chalcopyrite. Although more than 131 million tons of ore have been mined or identified in this deposit, it is still open to the east, and below about 2,900 meters.

Mine development of the GBT outcrop started in 1978, and the first ore reached the mill in December 1980. After yielding more than 62 million tons, GBT closed on October 31, 1994. The DOZ mine, now with 49.7 million tons of identified ore reserves, began producing in September 1988. We stopped work in the DOZ in April 1992, after pulling out just 2.4 million tons of ore, because there was so much ore coming out of Grasberg that the mill was at capacity. But work in the DOZ will resume in 1998 or 1999, when there will be some smiles at the mill. DOZ ore is dubbed "mill sweetener," because its already high average 1.72 percent grade can hit "hot zones" of fully 5 percent copper. Ore from this mine improves the overall quality of concentrate shipped.

Currently, the only underground production is coming from the IOZ ore body, which began to produce by block caving in January 1994, just after GBT closed down. The IOZ's remaining 21 million tons of ore lie in a thick band between 3,450 and 3,150 meters. With the IOZ mine, the Amole Adit, and numerous exploration and maintenance drifts being constructed, there is plenty of work going on below the surface. A total of 750 miners and mechanics report for work underground, about half of these in the IOZ.

Crucial to the success of Freeport's underground operations have been a group of as many as 330 Filipino miners, mostly from Baguio, on Luzon Island in the northern Philippines. Of these, only 84 are left, the rest having been replaced by trained Indonesians. We are gradually phasing out the remainder of the Filipino miners as Indonesians complete the process of gaining the required skills. No doubt, for many years, there will be a need for expatriate miners, but we believe that this will be for the purpose of introducing specialized mining equipment which will be required from time to time.

The IOZ is mined 24 hours a day, with anywhere from 100 to 120 men underground in production at any one time. Their efforts remove 6,000 metric tons of ore each day.

An underground mine requires a considerable amount of infrastructure before a single ton of ore can be removed. Obviously, one or more adits must be driven toward the ore body, and various drifts must be driven off these openings, including vertical shafts for ventilation and ore passes. Freeport's underground operations use seven truck-mounted, multi-boom "Development Jumbos." These machines, costing nearly half a million dollars each, drill the blast holes used to break the rock in the various horizontal mine openings or "drifts." "Long-hole Jumbos," of which there are six currently at work, are used to prepare blocks of ore for extraction and in the undercutting operation. The really big drilling machines work on the ore passes and ventilation raises. Two of these, owned and operated by a contractor, work in our underground operations. Freeport pays approximately $1,000 a vertical meter for their product: a hole 2–3 meters in diameter cut through very competent rock.

The electrically operated big-time drillers have recently been joined by "Srikandri," an Alpine Miner Road Header mechanized driller. We have been testing this piece of equipment in the work on the Amole Adit. The workers named this tough machine after one of Arjuna's wives in the popular Mahabharata stories. Srikandri is fierce, brave, and a master archer—in contrast to Arjuna's other wife, who is soft and feminine. This machine has the potential to excavate a tunnel more quickly and more cheaply than the other machinery in use, but the basic design of the Road Header was for coal mining, and it has seen very little use in a hard rock environment. The results so far have

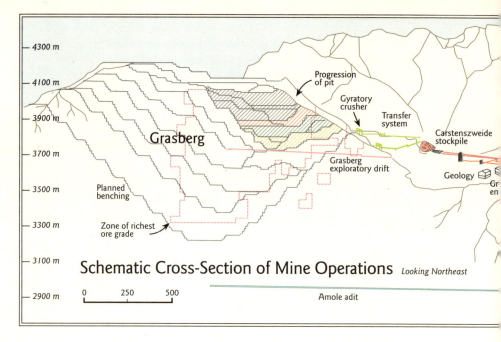

FIGURE 7.2

been mixed, but if it can be made to work, Srikandri will be worth its $3 million price tag.

Freeport's underground mine production operations today all use the block cave method, which is the cheapest and most efficient way to mine—or "stope"—the ore-bearing rock. In fact, a block cave mine can match the cost per ton of an open pit. In traditional underground mining methods, mine openings are driven on discrete levels throughout the ore body, and from those openings a finite block of ore is drilled and blasted. In the block caving method you only drill and blast out a horizontal band of five meters of rock, and the overlying ore caves under its own weight and fills that void. The caving process propagates upward, leaving us eventually with a large pile of broken up ore to extract. Our operating costs are low, because there's much less drilling, blasting, and labor required compared to the other underground mining methods. (See FIGURE 4.3 page 125.)

Block caving is used to mine ore bodies that are usually of a medium to low grade, and have a compact geometry. The mining method also requires rock of the appropriate strength. It has to be

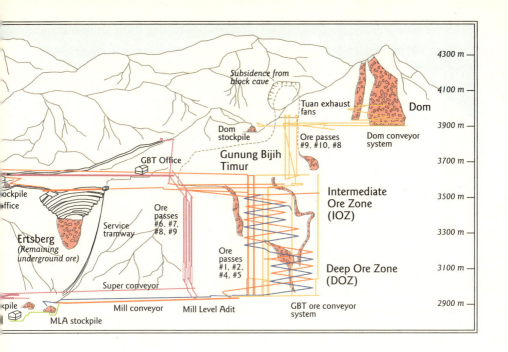

weak enough that when it is undercut it fails under its own weight with little or no additional blasting. It shouldn't be too weak, however, as the openings for hauling out the caved ore need to remain intact with a minimum of repairs. Block cave mining requires three separate levels of drifts: a top level for undercutting, a middle level, 18 meters below the undercut level, for receiving and extracting the caved ore, and a lower level for hauling out the broken ore on a system of conveyors.

At the top level, drill drifts are driven at 30 meter intervals from an opening into the ore body. Here teams of miners run Long-hole Jumbos to drill a series of closely spaced, 20-meter-long blast holes that fan out around the undercut drift. These holes allow the miners to place explosive for blasting the pillars, which will support the arch being undercut.

Blasting in our underground mine is carefully supervised. The area is cleared, guards are placed, and the location of all personnel are checked. A very large blast of pillars for an undercutting operation can require the evacuation of everyone in the mine to a safe location.

The explosives produce a dull, subdued thud, quite unlike the sound of an explosion in open air. The blast is followed by a rush of air sweeping in every direction through the open passages. After forcing the rock apart, the expanding gases produced by the explosion take the path of least resistance—the drifts and crosscuts.

Natural fractures in the rock are enlarged by the blasting, then gravity takes over as the back, or "roof," of a horizontal slice of rock fails, and it tumbles into the wide chamber blasted out beneath it. Once the process has begun, the cave naturally propagates upward at about a meter a day until it reaches the surface. When the caving process is complete, the broken rock can be extracted from below.

The network of extraction drifts, 18 meters below the undercut level, are excavated before the blasting takes place. The main passages on this level, called panel drifts, are lined with concrete, or a sprayed-on concrete called Shotcrete, and are located directly below the undercut drill drifts. A series of parallel extraction drifts, spaced 15 meters apart, runs from both sides of the panel drifts. These extraction drifts connect to what is called a drawpoint or chute, where the miners have access to the caved ore.

The caved ore is gathered from the drawpoints by diesel-powered load-haul-dump (LHD) machines, which are something like front-end loaders, hinged in the middle. Relative to the operator's position, these machines move crabwise, which solves the normal problem of favoring either forward or reverse. The 28 LHDs, together with 19 haul trucks, gather up the ore, much of it still boulder-sized. In block cave mining the processes of blasting and extracting the ore are consecutive, rather than concurrent, operations, which offers one of its biggest cost-savings over conventional mining. Once the ore is blasted, the drillers and blasting technicians can move on to another job, freeing up precious underground real estate and infrastructure for the loaders.

The loaders and trucks bring the caved ore to one of eight rock-breakers, large machines bearing a very powerful impact breaker, like a jackhammer, on an articulated arm. The rock-breakers break down the boulders until they can pass through an 18-inch grating to a bin and conveyor belt twenty meters below. The conveyor takes the rock to a crusher, which reduces it to 6 inches or less, and from there it moves to one of the ore passes, where it drops 300 to 550 meters, depending on the location, to the concentrator below.

THE CLUSTER OF buildings housing the milling and concentrating facilities sits at the head of the Aghawagong Valley, at an altitude of 2,850 meters. This is where the ore is crushed, the copper concentrated, and the resulting slurry sent down to the coast in a pipeline. The setting is drab, and most of the area is the grayish color of the ore. Mountain walls 800 meters high box in the concentrator, and space is at a premium here. The site is called "Mile 74," because it is 74 miles from the coast, and it is the end of the Freeport road. The aerial tramway begins in back of the stockpiles, and carries workers and equipment another 680 meters higher to the mine offices next to the old Ertsberg pit.

The two ore piles received from the mine are the lifeline of the concentrator, whose profitability and efficiency are based on a steady supply of ore. No ore, and concentrator operations have to slow. Other than slow-downs or stoppages in the ore supply, the concentrator's single biggest problems are caused by tramp material in the ore. Sometimes this is fairly innocuous—say some wooden beams and planks. Sometimes it is ore that is out of specifications for size, say 15-inch diameter boulders in an ore pile that is supposed to contain nothing larger than 7-inch.

But the biggest headaches are caused by pieces of steel debris: "How in the name of God can we get a seven-meter length of pipe, or two meters of metal mesh in there," goes a refrain. This kind of material can damage the crushers and belts, and is the major cause of downtime at the concentrator, which mill manager Charlie Wilmot estimates at more than an hour-and-a-half a day. There has to be a balance between the mine's ability to deliver the projected tonnage of material to the concentrator, and the mill's ability to handle anomalous tramp material.

The constant discovery of new reserves, and the increasing efficiency of the mine operations, has necessitated rapidly increasing production capacities at the mill. When mill superintendent Rachmat Pakpahan started working at Freeport in 1973, the mill was processing 6,800 metric tons of ore per day. By 1981 that figure grew to 9,500 tons per day, and by 1984, 12,500 tons per day. In 1988 the mill rate increased to 20,000 tons per day, and in 1990 jumped to 32,000 tons per day. This was all accomplished by increasing the output from the first concentrator building, the North Concentrator. "We maxed it out, and then some," said Pakpahan, who comes from North Suma-

The SAG mill is the heart of the new No. 3 Concentrator. This 14,200-horsepower machine efficiently and reliably reduces incoming ore to fine gravel.

tra. "But finally there was no way to increase our milling capacity—Grasberg ore was pouring down on us."

The second facility, the South Concentrator, was completed in 1991. The next year mill operations averaged 57,600 tons per day. Tuning the operations and increasing efficiencies where possible led to an average 62,300 tons per day for 1993, and finally 72,500 tons per day for 1994. Long before this milestone was reached, the decision had been made to expand the operation to 115,000 tons per day with a new 55,000+ tons per day SAG mill and concentrator north of the old facilities.

In 1993 we set June 1996 as our goal for completing the new facilities and reaching the 115K target. This would give us three years to train workers and sort the bugs out of the new equipment. A fast-track approach to design and construction eventually knocked a year off our schedule, and in June of 1995 the mill began processing at the 115,000 tons per day level. Today, given the continued good performance of the installed equipment, we expect the long-term average to be 125,000 tons per day.

Ball mills running in the South Concentrator. Our original concentrator facilities have increased in capacity from 6,800 tons a day to more than 70,000 tons a day.

"We take the normal production, and push one step further," Wilmot says. "If we have to slow down because of a lack of ore, we'll make it up. Do we need another crusher? Another 30 workers? We figure it out."

Once there was a major ore pass failure. To correct this problem, supervisors aggressively and closely managed the critical work on the ore pass replacement and conveyor construction, bring production back to normal as quickly as possible. Then they pushed the entire system at every part of the operation until production made up for the losses.

"If it's economic, we do it," Wilmot says. "The company has the foresight to give us the capital we need. We run hard. We use our equipment to the maximum—and further, if it can be done safely. That's good business, the maximum utilization of your capital. It might mean a mess, running the equipment at the edge. But our engineers figure out a way to clean it up. What counts at the end of the day are the 'rocks in the box.'"

The semiautogenous grinding mill, and the No. 3 Concentrator that it feeds, is the new pride of the mill facility. This gigantic piece of equipment works by forcing the rock to crush and grind itself down to a manageable size. The SAG mill is better for handling our ores than the older screens and crushers feeding the North and South Concentrators, and currently processes about half the ore reaching the concentrators. The SAG mill is also simpler to operate than the older ore-crushing and screening facilities, which are notoriously difficult to maintain.

The biggest problem for the new concentrator was that there was simply no more level ground at the mill site. Over a thirty-month period, the Amole Ridge was leveled to provide 8.5 hectares for the new facilities. The area does experience occasional earthquakes, and the fill and buildings were designed with seismic safety in mind—we have never experienced an earthquake-related loss in our entire operating history. The SAG mill concentrator cost $300 million, with another $100 million for the power plant to run it. Perhaps 40 percent of this cost went for excavation and site preparation. The No. 3 Concentrator started operation in February 1995, just 18 months after foundation work began, which must be some kind of a record.

The SAG mill is a monster, a huge, 34-foot diameter barrel that slowly turns round and round. The ore, which comes directly from the Amole stockpile, gives off a low grumble from the inside as it grinds itself to bits, assisted by steel balls up to 4 inches in diameter. To spin at its leisurely 10.6 revolutions per minute, the SAG mill requires 10.6 megawatts (14,200 horsepower). In go cobbles and small boulders; out comes nothing coarser than fine gravel. Ore that makes it through the SAG mill's 8-millimeter slot goes on to the ball mills, units that use 3-inch diameter steel balls to grind the ore to the consistency of fine clay.

The SAG and associated ball mills create a dull roar in the east end of the No. 3 Concentrator building. A steel mesh walkway, 20 meters above the floor, runs from the SAG, past the vibrating screens in front of it, and above the two ball mills. The control room offers relief from the noise and vibration. All aspects of the operation are monitored through state-of-the-art sensors and video screens, which dance on specially designed vibration-isolating pedestals.

Everything is individually measured: air flow, depth of froth, the flow of slurry. This data is then fed into a computerized expert system, which adjusts the settings of the equipment. This system has significantly improved the grinding quality at the mill, and our installation is the envy of the industry.

The powdered ore is concentrated in a process called "froth flotation." The finely ground rock is mixed with water, to which lime and traces of reagents are added, and air is bubbled through the mix. The lime raises the pH of the slurry, and through the action of the reagents, the ore minerals become "air avid," fastening themselves to the surface of the bubbles. These bubbles, with the copper, gold and silver clinging to them, are concentrated in a frothy layer at the top of the flotation cells, and skimmed off. The entire process, from raw ore to concentrate, takes about an hour. (See also Chapter 9, page 251.)

Froth flotation technology was developed in the first decade of this century when scientists discovered that metals were either hydrophyllic (they "like" water) or hydrophobic (they don't). Metals in this latter category can usefully be concentrated using flotation. Chalcopyrite is the principal copper-bearing mineral in Grasberg ore, about 90 percent of the mill's supply, and bornite is the principal copper-bearing mineral in the underground skarn deposits. Two other valuable minerals found in lesser quantities in both ores are chalcocite and covellite. All of these copper-bearing minerals are hydrophobic, and will adhere to the air bubbles and leave the flotation cells as copper concentrate. Silicate, calcite and magnetite are hydrophyllic and do not float in the flotation cells. These are discharged in the muddy waste called "tailings."

Although froth flotation is essentially a physical, rather than a chemical process, three chemicals are added during the concentration process. Lime makes the solution more basic, raising the pH from the natural 8.0 to a quite alkaline 10.5. This requires about two kilograms of lime for every ton of ore processed. While lime is now being imported, limestone deposits have been identified in the mine area, and within two years the mill should have a local source. Small amounts of two reagents are also added, which aids foam formation and the adherence of the copper particles to the air bubbles. Most of the reagents, 95 percent, stay with the concentrate, and the small amount discharged with the tailings is not toxic.

Up close, one can see a layer of golden-brown bubbles forming at the top of the collector cells, and overflowing the edge. This is the heart of the process. In additional rows of cells, the "cleaners," the minerals are further concentrated, and 50-foot-tall cleaner columns

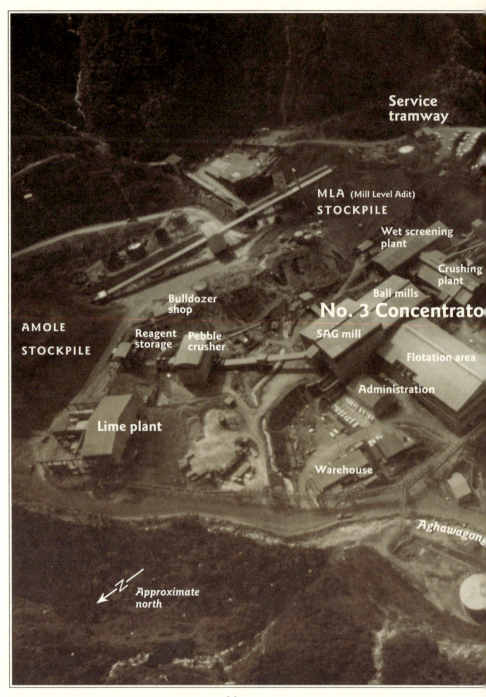

Map 7.1

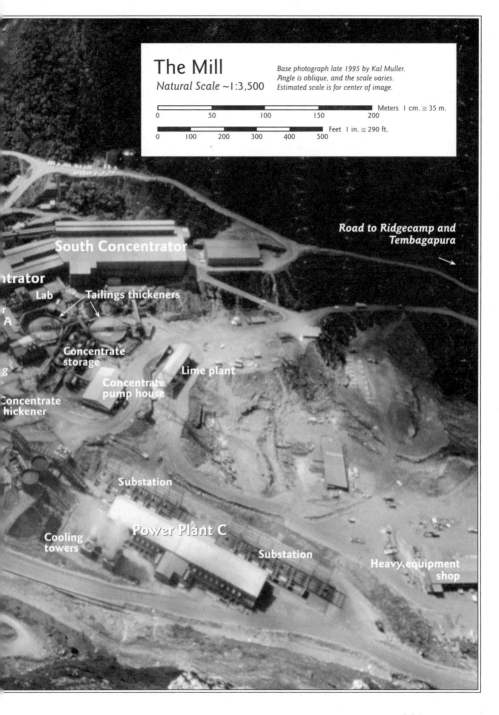

perform the final concentration. Instruments inside the pipes, tubes, columns, and tanks relay information to the control room, and any necessary changes in water or air pressure are made by issuing a few instructions to a computer. How fine the ore is ground, its time in the flotation cells, and the exact amount of reagents used all affect the recovery rate, and Freeport engineers are constantly experimenting with and fine-tuning these variables.

Water is the essential component of the concentrating process, and the mill requires a lot of it—20,400 gallons a minute, 24 hours a day. This works out to .81 tons of water for every ton of ore processed. In such a wet, foggy area, where rainfall averages 8 millimeters a day, water supply would seem to be the least of the mill's problems, but during occasional short periods of drought water can become a scarce commodity. For a brief period in 1991, mill production had to be scaled back 20–25 percent because of a shortage of water, an expensive situation that nobody wants to repeat—and probably will not have to, as a result of the measures taken to bring additional water to the concentrators.

The mill tries to capture all the possible sources of water in the area. The Aghawagong River, the closest obvious source, can supply just a bit over 10 percent of the mill's needs. Lake Wilson, the water-filled former Ertsberg pit, provides about the same amount, and its 20 million metric tons of water serves as a reserve. Although 20 million tons of water seems like a lot, the mill's requirements are so great that this reserve will only last 11 days.

Whatever rain falls in the area runs off very rapidly because of the steep terrain and sparse alluvial cover. So four low dams, or weirs, have been built, to collect all this wasted run-off. Trash and debris is also swept down into these weirs, and they are cleaned daily. Together, the Macken Dam, the Hasan Dam, the Carpentry Shop Dam, and a dam collecting the drainage from the HEAT road provide perhaps 20 percent of the mill's daily needs.

The single biggest source of the mill's water is the Amole Adit, which alone meets more than 20 percent of the mill's current needs. The MLA ("Mill-Level Adit") to the GBT deposits also brings a substantial amount of water. These adits were not primarily designed to provide water, but as production has climbed, this function has become very important.

Water is reclaimed from the tailings by means of tailings thickeners or settling tanks. Clean water is decanted from the top, and the thickened tailings are discharged from the bottom. In reclaiming water from the tailings, we are also able to reclaim much of the lime and reagents as well. In times of normal rainfall, we have enough water to process more than 130,000 tons of ore per day. The now advancing Amole Adit will provide added water, and if need be, we could construct more tailings thickeners. Mill engineers have, by necessity, become very frugal with their water, and recycle or reduce their water needs wherever possible.

The mill also requires a great deal of electrical power, the generation of which is another major endeavor. The power plants feeding the mill produce an average of 70 megawatts an hour, enough to supply 50,000 homes—televisions, microwave ovens and all. Generating this much electricity requires 216,000 gallons of diesel fuel every day.

All the water, electricity, concentrator machinery, and the work of thousands of skilled men have just one aim: to extract the most copper, gold and silver out of the ore. Although it is physically impossible to get it all out, our operators are proud of Freeport's recovery rate. In the big Chile mines, 80 percent is considered good. Our 85–90 percent recovery is exceptional.

Counting both full-time staff and hourly employees, 844 men work at the concentrator. The full-time staff totals 109, of whom 84 are Indonesian and 25 expatriates. Most of the more highly skilled workers are engineers—mechanical, electrical, metallurgical and instrumentation. An in-house accounting section handles the finances. Communications at the mill are in both English and Indonesian. The expatriates have been training Indonesians for two years now in the operation of the new facilities.

The consensus at Freeport is that the mill has a first-rate training program, and the best career and advancement programs of the operations in Irian Jaya. These programs start at the beginning of employment. When the management became aware of language deficiencies among the 164 Irianese workers at the mill, most of whom lacked formal schooling, a special "bridge-training" program was created for them. They were taught on the job, three times a week. At the end of the course, they received a primary school certificate. "Now you can see the Irianese with a technical manual—they're reading it," Pakpahan says. "That makes me feel good."

Concentrated slurry can be seen as a shiny froth on the top of this column cleaner flotation cell. A strong stream of fine air bubbles is the key to the process.

The finished product of the mill is enriched slurry, containing 30–34 percent copper, and about .9 ounce of gold per dry ton. When it leaves the mill, the slurry is about two-thirds solids and one-third water, with the concentrate being made up of extremely fine particles of enriched ore. At a production rate of 115,000 tons of ore a day, the daily output from the mill is 4,100 tons of concentrate. The Freeport mine concentrate is much sought after by smelters, because it contains very low concentrations of such deleterious elements as bismuth, fluorine, and arsenic.

The slurry travels the 109 kilometers from the mill to port site at Amamapare through steel pipelines. Its pace is a leisurely 3.5 miles per hour, so the journey takes about 20 hours. These crucial pipelines, which parallel the access road, fall under the special care of Ron Walter. When Walter first started at Freeport in 1988, the slurry line was a single, 4.25-inch pipe. This was converted to deliver diesel to the mine and mill, and two 5-inch lines were laid for concentrate. Now one of the 5-inch lines is being mothballed, and two additional 6-inch slurry lines have been laid.

An ore ship being loaded at Amamapare. The site, about 6 kilometers from the sea at the junction of the Aikwa and Jaramaya Rivers, is as far inland as the ships can go.

Walter keeps several small sections of slurry pipe in his office. One is worn through in a section to where it is egg-shell thin, and another is full of hairline cracks. The pipe is made of high tensile strength steel, and if it is operated correctly, is almost impervious to wear. The key is to manage the volume and consistency of the material in the pipe. If the slurry is too thin, particulate matter will settle out and drag on the bottom of the pipe, accelerating its wear. And the system works best at design volume. Basically, you need to pump at the calculated rate to minimize the problems you could have.

The greatest normal wear occurs during periods when we have emptied the feed tanks and are flushing the line with water. During this period, the dynamics of the pipeline change, and the flow in a portion of the line accelerates and wear occurs along the bottom of the line. An eight-mile stretch of pipe is affected by this problem, and we keep the wear to acceptable levels by rotating the line a quarter turn every year or two. Fortunately, in this problem area of the run, the pipe is under negative pressure, and if a hole should wear through, it would suck air, but it would not leak slurry.

The first pipeline was exposed, and in the 1970s, it was cut in protest by members of the Organisasi Papua Merdeka, the small guerrilla group opposing Indonesian rule of Irian Jaya. "Apparently all they had were hacksaw blades," Walter says, with a certain unavoidable respect for the amount of work such a task must have involved. "They had enough manpower to cut out several hundred feet of pipe, which disappeared over a cliff."

At port site, the slurry is filtered, dried, and warehoused. Close to 100 ore-carrying ships call at port site each year, ranging in capacity from 5,000 tons to 35,000 tons. The drying operation is closely monitored so that the concentrate contains 7–9 percent water. Any less moisture and the material becomes too dusty; any more and it becomes semi-fluid, making the ship dangerously unstable.

CHAPTER EIGHT

Exploration

*Plate tectonics—A plumbing system—The New Guinea mineral trend—
Aeromag—Hoisting—Wabu Ridge and the 'Fighting Irish'—Etna Bay—
Jurassic Park—Life in the field—Some ugly rock—Drilling*

FREEPORT'S GEOLOGISTS ARE currently looking for signs of mineralization in a vast patchwork of contract areas spread across 7.3 million acres of Irian Jaya. So far, we have found more than seventy "anomalies," interesting geological inconsistencies that could point to the presence of mineable ore. Finding these requires poring over satellite and radar images, commissioning and then studying aeromagnetic surveys, hoisting down from hovering helicopters, and simple, old-fashioned hikes up streambeds with a geology pick, notebook, and sample bag.

The history of the copper, gold and silver deposits sought by Freeport goes back a long way. The basic elements have been there since the earth formed. But the creation of New Guinea, and the geologically active mountain range that presents such a rich mineral prospect, is more recent. The world's continents, including New Guinea and Australia, were not always where they are today. According to the theory of Plate Tectonics, by 250 million years ago they had all come together into a single, large land mass called Pangaea, which can be seen in the near perfect fit of South America and Africa, polar North America and polar Asia, and Australia and Antarctica. Pangaea, during the Mesozoic Era (65–240 million years ago), first split into two great continents: Laurasia, which included present-day North America and Eurasia; and Gondwanaland, which included present-day South America, Africa, Antarctica, India and Australia. The body of water that opened up between them as they swung slowly apart is called the Tethys Sea.

As Gondwanaland split from Laurasia, it began itself to break up into what would become the familiar southern hemisphere conti-

Old drill and helicopter pads on the Wabu Ridge. Exploration is an expensive and often frustrating process, and the huge skarn body here is still not fully understood.

nents of today. At first India and Australia headed north separately, but by the time India had made about half the journey from Antarctica to Asia, about 60 million years ago, the crustal plates upon which these continents moved welded together into what we now call the Australian Plate. This plate, a "raft" of cool, crustal rock sliding on the molten magma beneath, moved northward at a rate of 200 kilometers every million years—the highest speed of any drifting continent. A few tens of millions of years later, India, the westernmost edge of the plate, "collided" with the Asian mainland. The tremendous force of this meeting raised the Himalayan plateau, substantially altered the geology of almost all of China, and squeezed mainland Southeast Asia—Burma, Thailand, Vietnam, Malaysia—off to the southeast. This "collision," although the rate of movement has slowed to one-fourth of its earlier speed, is still going on.

In the east, the leading edge of the Australian Plate met no great continent with which to collide. Instead, it has steadily been subducted beneath the Pacific Plate. When two plates meet in an area where there is no great thickness of crust providing resistance, the

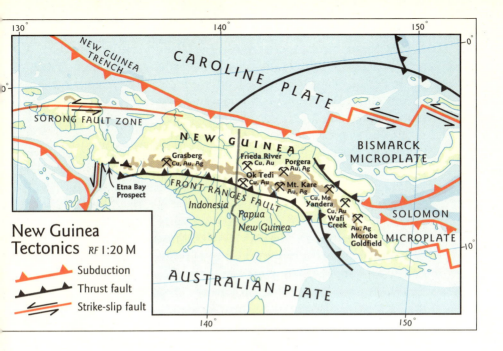

MAP 8.1

edge of one slides under the other, where it is driven down to the liquid mantle rock in a process called subduction. As the edge of the Australian Plate has been driven beneath the Pacific Plate, the oceanic crust has melted, creating magma that rises to the surface. The long arc of volcanic islands that includes the Andaman Islands and much of Indonesia—Sumatra, Java, the Lesser Sundas, and the small islands of the Banda Sea—was formed this way.

At the far eastern end of the Australian Plate, where the continent of Australia rides, the situation is quite different. New Guinea is really just the leading edge of Australia, and the Arafura Sea that makes New Guinea a separate island today is so shallow that just 15,000 years ago the two were connected by land. Most of the rock making up the mountains of Irian Jaya first formed 170 million years ago in the Jurassic Period, when Australia and New Guinea were still part of Pangaea. Weathering of this rock eventually built up a layer of sedimentary deposits 3,500 meters thick. When this substantial package of rock reached the subduction zone at the edge of the Pacific Plate, it jammed the trench, buckling its own sedimentary rock 4,000 meters high to create the southern edge of the island's central cordillera. As pressure continued to build, the marine sediments forming the edge of the Pacific plate also buckled and rose, creating the northern edge of New Guinea's cordillera of mountains. Weathering of these mountains created the great, flat Meervlakte, or "Plain of Lakes" to the north, and the wide alluvial plain of the south.

Today, the plate boundary around New Guinea has become very complex. The Pacific Plate has broken up into the small Caroline Plate and the Solomon Microplate. Some tectonicists believe that the subduction zone here is in the process of reversing, that New Guinea has proved such a stubborn block that the Pacific Plate will begin to slide under the Australian Plate at New Guinea. If this happens, New Guinea will begin drifting through the Pacific, consuming the islands in its path. (See MAP 8.1, page 217.)

The tremendous geological activity of New Guinea is what makes it an attractive place to look for minerals. As the plates met and the great mountains formed, cracks and fissures opened up, into which igneous material, under great pressure from all the activity, rose and cooled. Sometimes the material had sufficient energy to break the surface, forming a volcano. These intrusions, as they are called by geologists, are the key to finding gold and copper.

By itself, an "intrusive anomaly" is not enough to guarantee a rich deposit of copper and gold. It is in the chemical interaction between a high-energy igneous intrusion and existing sedimentary deposits that ore is formed. Although the relatively recent collision of the Pacific and Australian plates has much to do with the size and structure of New Guinea's mountains, the limestone bedrock of which they are made has a much older history. The oldest sedimentary layer is the Kembalangan Formation, which dates from 136 to 65 million years ago, and is composed of a thick sequence of siltstones, sandstones and quartzites. More recent—46–53 million years old—are the Waripi, Faumai and Kais Formations, which overlay the Kembalangan. The limestone of these more recent formations provides the appropriate host rock for the ore bodies sought by Freeport's exploration geologists.

Our currently operating mines are all in a relatively small area, rich with ore bodies. The Grasberg and Ertsberg intrusions lie above a deep intersection of faults, some 10 to 15 kilometers underground, which have provided the necessary pathways for the magma to reach to mineable depths. Geologists call this system of faults and fractures a "plumbing system," which directs the fluid rock from the mantle all the way up to the surface. (See MAP 8.2, page 225.)

An igneous intrusion sets off a number of chemical events. As the liquid rock reaches a chamber, it begins to cool and crystallize. But the copper and gold do not settle out, instead remaining in the hot fluid. Thus the fluid part of the magma is relatively enriched in copper and gold, and the solid portion relatively impoverished. This very high-energy fluid rock, mixed with other volatiles like superheated water, continues to stream upward, either through the already cooled part of the intrusion, or through other passages in the plumbing system. The very high activity of these fluids causes thermal and hydrothermal alteration of the rock in its path.

This is not a one-time event. The crystallization chokes the fissures and they plug, and the fluids find another outlet. Then, perhaps, the intrusion cools and its activity ceases, and another intrusion comes along, using the same plumbing system. Eventually these repeated processes form a concentration of minerals that may be of economic interest. A "concentration" means less than 1 percent copper for most ore bodies, and 1–2 percent for a very rich deposit. Since the earth's crust averages .03 percent copper, a normal deposit represents about a 33× concentration.

There are many mountainous areas on earth with no valuable minerals. But ever since 1878, when a small gold rush swept the Laloki River, about 10 miles inland from Port Moresby on the Papua New Guinea side, it was known that at least some of New Guinea's mountains had gold. This first gold rush was short-lived, although a few Australian prospectors worked the eastern half of the island in the late 19th century. The big rush came in 1926, a few years after William "Shark-eye" Park struck gold on Koranga Creek, a tributary of the Bulolo River. In the decade of the 1920s, £1.3 million in gold left New Guinea; in the 1930s, this rose to £15 million.

On the Dutch side of the island, early geological explorations concentrated on the petroleum deposits near Sorong, on the western tip of the island. Jacques Dozy, when he identified Ertsberg and Grasberg, was working in Dutch New Guinea as an oil geologist. When Freeport began our mining operations in the early 1970s, both Kennecott and Newmont explored sections of the highlands. They found two significant intrusives near Enarotali, but because of economic and logistical considerations, never fully explored them. During the 1970s, Freeport's own exploration geologists were busy discovering Gunung Bijih Timur and the other rich deposits near Ertsberg. Then copper prices dropped, and the exploration program waned.

Geologists have for some time been aware that the igneous rocks most likely to support copper-gold porphyry mineralization appear to be 2–3 million-year-old intrusives of the late Pliocene Epoch. Recent knowledge about low-grade copper porphyry deposits led to increased production first in the southwestern United States, then in Chile, where the world's largest copper mines operate today. As the geology of high-grade porphyry systems began to be understood in the 1960s and 1970s, big copper mines opened on the Papua New Guinea side of the island, including Bougainville, on an island near the Solomons, and Ok Tedi. The Pacific mines brought too much metal to the marketplace, and prices dropped drastically. Eventually, demand caught up with supply and prices started rising again.

EXPLORATION GEOLOGISTS IN a place like Irian Jaya lead strenuous and adventurous lives. Work requirements include a love of the outdoors, a tolerance for leeches, mosquitoes and other unsavory creatures, and the ability to trek for hours on end in rivers and dense, untrod vegetation. One can't be too in love with luxury, either, as geologists spend weeks at a time in very basic camps: a tent to sleep in, a hole in the ground for a toilet, and a creek to wash up in.

Dave Potter is the chief of our exploration program. He spends a lot of time in the office these days, and you can usually tell when he needs to get out to the field when he starts fidgeting with his geology pick, which he keeps always at hand. Potter has nothing but praise for his geologists, particularly the Indonesians. "The Indonesian geologists are better mineralogists than the expats," he says simply. What they lack is experience. Most of the expatriate geologists working for us have experience at dozens of other sites, which is what makes them valuable. The Indonesians, many of whom are young and fresh out of university, have field experience in only a few places. Potter has set up a program to bring the Indonesian "geos" to places like Mexico, where Freeport has a small concession, and Australia—both dry and unforested, very unlike Irian Jaya. "We basically work by analogy," he says. "You walk over some rocks, your eyes see them and your mind digs into the files and says 'You've seen the same pattern of

The mule of our exploration program. Conditions in Irian push the limits of helicopter design, and this Bell 205 has modified blades to raise its operating ceiling.

rocks at Ok Tedi, or Escondida.' You can't do that if you've never been to any of those places."

Large-scale exploration geology, such as Freeport practices in Irian Jaya, is a very expensive and time-consuming project. Although exploring so many square miles anywhere would be difficult, Potter says, "The terrain and lack of infrastructure support is what makes Irian a bitch." Only the Amazon, or the Antarctic, would be as bad.

A geologist does not immediately start scouring the streambeds and outcrops for signs of mineralization. Long before any groundwork is done, Freeport geologists use their knowledge of general trends and broad geologic patterns to come up with a hypothesis as to just which areas might be good prospects. They conduct literature searches, to see if past geologists have noted anything relevant. If the general patterns and written material allows for hope, and the company is ready to commit resources, a specific area is chosen.

The first contract of work area outside the immediate area of the original Ertsberg mine basically covered all of Irian Jaya's highlands. We picked the highlands based on a "trendology" model. The same package of rocks, of the same age, are found on both sides of the border. On the Papua New Guinea side, there are more than twenty active mines. On the Irian Jaya side, just our mine. We were certain there must be more mineable ore in there somewhere.

As Jim Bob Moffett says, "We looked at the Papua New Guinea side and the Irian Jaya side and said, 'Wait a minute, this thing should be a mirror image trend.' Mother Nature doesn't know anything about political boundaries, she didn't know she was going to be cut in half."

Copper porphyries found in the same part of the world tend to have a characteristic composition, which will be common to many of the deposits in a general area. For example, the copper deposits found in the region extending from South America into North America contain important concentrations of molybdenum, but although they have some silver, usually contain very little gold. In contrast, deposits found in the region of the South Pacific from Indonesia into the Philippines contain very little molybdenum, but important concentrations of gold and silver. Of the 42 copper-gold porphyries that so far have been identified in the world, 31 of these can be found in an area of the globe that encompasses the Philippines, Indonesia, and the northern part of Australia. We have been quite fortunate with our

Grasberg and Ertsberg complex of ores to have an important precious metal byproduct, in addition to an excellent copper grade. Others can mine the molybdenum.

This mirror-image trend gets you started, but there is a whole set of parameters that must be in place if you are going to have a mineral occurrence. You need the right stratigraphic package, and you need an intrusion. Basically, this would apply to almost all of Irian Jaya's highlands. But this might not be enough. At Grasberg, for example, cross-structures, major regional cross-structures, are an important part of the geology. The western end of our contract of work has a lot more cross-faulting than the eastern end. Looking at satellite data—U.S. Landsat, the higher resolution French SPOT, and satellite-based radar images—we can pick key intersection points where the general east–west run of the mountains is broken by cross structures. Beneath these points are very old basement structures that have been reactivated over time. This provides the source, and the "plumbing system," for mineable ore deposits. (See MAP 8.2, page 225.)

In other parts of the world, after looking at geological maps and satellite images, a crew would go out and have a look at the ground before acquiring rights to the land. In Indonesia, however, a company must sign a contract of work before it can send out a crew. This COW provides access to the mineral rights for a period as long as 30 years, with the possibility of two 10-year extensions, or as little as 3 months. The COW includes a procedure to free up the rights to land as it proves barren of useful minerals. A minimum of 75 percent of the contract area must be released within eight years. The remaining 25 percent, if economic mineral deposits are found, can be designated a mining area. Only at this point is the company able to secure surface rights, as needed, to facilitate access and to develop and built mine and mill facilities.

Currently, we are operating under two separate COWs with the Indonesian government, totaling 7.3 million acres of land. In 1991, we signed a 30-year contract to explore Block A, a 24,700-acre area around Grasberg and the original Ertsberg mine, and Block B, a 4.8-million-acre area spread through the highlands all the way to the Papua New Guinea border. This contract includes the possibility of two 10-year extensions. In 1994, under the auspices of an affiliated company called P.T. IRJA Eastern Minerals Corporation, we signed a contract for rights on three additional blocks of land—one around

Field geologist Philip Petrie and an Indonesian trainee do some homework by visiting the Kembalangan type formation, on the north coast of Lakahia Bay.

Etna Bay, one around the headwaters of the Tariku (Rouffaer) River, and one around the Kembu River, just southwest of Dabra—totalling 2.5 million acres. The contract for these is also 30 years, and includes two potential 10-year extensions.

Once the legal rights are secured, teams of field geologists are sent out to do preliminary groundwork, and the company conducts remote sensing fly-overs. This process, called "aeromag," consists of flying fixed-wing aircraft and helicopters outfitted with sensitive equipment to measure magnetic emissions along a precise grid over the land being surveyed. The instruments detect small variations in the magnetic field of the surface rock, which yields clues as to the structure and composition of the rock. Aeromag is expensive, costing $30–$50 a kilometer, but, especially in a place like Irian Jaya, where access to the ground is so difficult, it can save a tremendous amount of time and money compared to sending in geologists. In one of the first exploration blocks, permit holdups kept us from being able to fly magnetics before sending out ground teams. Where an aeromag survey can cost $1 million, Potter estimates that, in this case, the extra

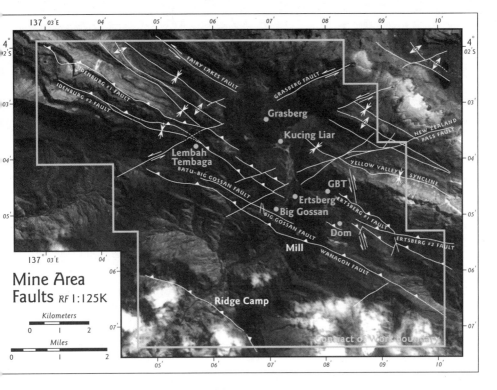

MAP 8.2

groundwork that had to be done because there were no clear magnetic targets took a year—and cost about $25 million.

Aeromag data is an important part of the equation, but not the only one. Intrusive or volcanic rock is often highly magnetic, and porphyry copper is an intrusive. Gold is also sometimes associated with magnetite, a highly magnetic form of iron. The magnetic data points geologists to the heat source, the center of the ore-creating system—what our geologists call "the bull's eye." Grasberg, for example, has a very magnetic mineralization phase and appears as a huge magnetic anomaly. At the same time some porphyry copper deposits aren't magnetic. In these, hydrothermal alteration has destroyed the original magnetite and, though they might be rich in copper and gold, you will see at best only a weak magnetic anomaly. The most reliable source of information is what geologists call the "ground truth." This can only come from someone who has sampled the rocks, and viewed and mapped the geological structures in the area.

Freeport's exploration areas in Irian Jaya are rugged, forested and served by not a single road. Since we began our expanded exploration program in 1991, some 95 percent of the time our geologists are dropped in by helicopter. The usual procedure would be to send a ground team in to hack out a clearing for a helicopter to land, and then run the exploration program from this site. The problem with this method is that it takes time, costs money, and causes some environmental damage. Instead, we have perfected a system of hoisting. Instead of landing the helicopter, our men hoist down—and back up—from a hovering chopper. No other company has used hoisting to the extent that Freeport has, and some of our geologists have more than a thousand hoists under their belts.

This operation is run with military precision. Safety is the paramount concern. We have a policy that any member of the team, for any reason, can call off a hoist at any time with no recriminations. Even if it's just ten o'clock in the morning, if the wind seems a bit tricky, if the hoist master is having a bad day, if the site just doesn't look good, anyone can call it. Nobody goes down that long, steel cable unless everyone on the team feels right about it.

The geologists who drop from the helicopter are performing what we call "junction sampling." They always land on a stream bed. In Irian Jaya a stream is an ideal place to analyze rock because, with the area's high rainfall, everything washes into the streams. Also, in such a heavily forested region, a stream bed is one of the few places where the underlying rock is even visible. The geologists gather a representative sample of gravel from the stream, called "float," and then pan a sample to concentrate any gold and other heavier material. These samples, because they have to go back up the line to the helicopter, are quite small. The basic test at this stage is the presence of intrusive float in the streams, and gold counts in the pan. As one of our senior geologists says, at this point they "look for the smoke and don't worry about the fire."

These samples are carefully labeled, and sent to the lab in Timika where they are "dressed," dried and ground, and sent off to laboratories in Jakarta, Balikpapan—a city in the Indonesian province of South Kalimantan—and the United States and Canada for assaying. The arrival of the lab results begins the period of number crunching and interpretation.

The follow-up phase to the initial reconnaissance work will only occur if the most senior geologists and the company management are fairly certain that there's fire under the smoke. This is the stage at which budgets disappear quickly. Rough wooden helicopter landing pads are built and camps are set up, usually in the clearings provided by streams. Then the geologists come in and develop a detailed map of the area, and examine the rock more closely. They conduct traverses of all the streams, and note, mark, measure and sample any important geological features they encounter.

Freeport's Indonesian and expatriate field geologists work in small teams, together with local Irianese assistants. These assistants are extremely helpful. They are familiar with the local language, know the local topography, and are strong and sure-footed in the demanding terrain. Senior geologist Peter Doyle recalls working in the mountainous area around Bilogai and being pulled up waterfalls by his assistant Tom, a Moni man from Sugapa. "He's carrying the rock samples, the front of the measuring chain, and the *parang*, and he's standing there in his bare feet, holding on by one arm. And he reaches down with his free arm and pulls up my eighty-five to one-hundred kilos."

Field geologists have to be in top physical shape, and very capable at negotiating steep and rugged terrain. Breaking a leg in the middle of Irian Jaya is not the same as breaking a leg in a public park in the United States. Geologist Gary O'Connor is rather laconic when asked about the hazards of working in the field. "Close calls? Not really. I've had trees fall on me. I nearly had my neck broken two or three times."

When several geologists are working the same area, a good-natured rivalry often develops. It's much more satisfying to be bringing back samples containing high mineralization than to be bringing back barren rocks. In the early days at the Wabu prospect, north across the mountain range from Grasberg, Doyle and O'Connor worked together as one exploration team, and geologists Hamish Campbell and Doug Mackenzie worked together as another. Doyle tells the story:

> We started up the big drainage at the west end, taking tributaries as they came. And as we went, we were getting interesting skarn, altered intrusives. We were climbing waterfalls and there was a lot of limestone. When we got to a junction about two-thirds of the way up, we said to Hamish and Doug, 'Okay, you guys go up the main stream.' At this stage the main stream was still fairly substantial. Gary and I took the side stream.

GREG PROBST, CIRCA 1993

Hoisting from a hovering helicopter is the most efficient way for a geologist to get into Irian's rugged interior. No other company uses the technique as extensively.

We went up the smaller drainage and kept finding skarn, and really good mineralization. At that point we had about the highest assay samples, I think it was 36 or 26 grams of gold in a float sample. And every day for the next week we came back with mineralization and interesting rocks. And when you're hitting, there's nothing better. It really helps the morale. Doug and Hamish went about a hundred meters beyond the junction, hit sediments, and never saw an interesting rock after that. So we called them the 'Limestone Cowboys' because that's all they were finding, just dead-looking rock. The flying Scotsmen were finding nothing. We named our section the 'Fighting Irish Zone' for O'Connor and Doyle. It's since been shortened to FI.

Once the area is mapped, sampled and detailed, it is time to send down some drill holes to see what's really there. "Drillers," Potter says, "are the lie detectors that we geologists live in fear of." A geologist might have what he thinks is a great target, he says, but after a few holes it can turn into a "barking dog"—or, of course, a great success. Doyle calls drilling the "acid test of any geological theory."

A field geologist can only sample whatever rock pokes through the thick vegetation, or washes down the waterways. This, combined with satellite and aeromagnetic data, allows him to come up with a

The hoisting operations are run with military precision, and if any participant—hoist master, pilot, or geologist—feels conditions are not right, the hoist is scrubbed.

rough idea of what lies under the surface. But this is still just a hypothesis. Samples from deep in the earth must confirm the surface data. Valuable mineral deposits lie underground, in three-dimensional bodies which vary tremendously in size and shape. And the mineralization percentages are far from homogeneous within any potential ore body. So samples must be brought up and analyzed.

The only way to do this is by using diamond drills to bring up a long series of core samples. The length and angle of each drill hole is based on a geologist's educated guess—and it had better be a good one, as just getting the core out of the ground and assayed costs $100–$200 a meter.

The location and angle of the drill is crucial, as a meter or two up or down or sideways can mean a marked difference in the rock's mineral content. After the initial drilling in any location, the results help to determine if more holes are needed, to what depth and at what angle. Eventually, a program of core sampling produces a three-dimensional map, a cross-section of what lies below. At Wabu, a large, complicated skarn body, this program has so far produced 63 drill sites, 144 drill holes, and more than 55 kilometers of core samples.

When the core comes up, it is marked and carefully laid out in a flat tray. The trays are hauled to the nearest chopper pad, where they are flown to the core shed. There, each tray is photographed, as insurance in case it is later jumbled or lost. If the ore body should eventually be worthy of mining, the photographs can also give a mining engineer a quick overview of the stability of the rock. The cores are also examined by a geotechnician, who measures the fractures and specific gravity of the rock, and decides which core intervals he wants sampled by the lab. This detailed sampling looks for lithologic boundaries, general mineralization, alteration and areas of increased vein intensity. The samples sent out for analysis are split—half remains in the core box, and the other half goes to the prep lab, and from there to the laboratory for analysis.

A typical drill rig is run by two Indonesian drillers and four local helpers, taking turns on 12-hour shifts. At a normal location, five or six holes are drilled, at different angles, going down to 400–500 meters. The Wabu prospect, near the village of Bilogai, taught everyone the practical aspects of drilling in the field in Irian. It was not easy going. Freeport was beginning to use a new type of drill, a lightweight, hydraulic rig called the Longyear LF 70.

Some of our expatriate drillers, accustomed to the heavier and more traditional mechanical drive Longyears, were reluctant to use the new rig to the depths required. Fred Elliott, a very experienced driller who heads our operations out of Jakarta, flew out to Bilogai. He had gone over the specifications of the LF 70 very carefully, and he knew it was possible to drill to the depths the geologists wanted.

"The expats said it couldn't be done," he said. "I told 'em: 'There's the plane. If you can't do it, you're fired, and I'll do it myself.' It got done. The Indonesians thought the machine was fantastic, but some of the expat drillers, you have to hit them on the side of the head to bring them around."

Soon after Elliot's pep talk, we began getting records from these new rigs. With a size H bit, 89 millimeters, the drillers reached 650 meters. For a size N, 70 millimeters, 917 meters. For a size B, 65 millimeters, 1,001 meters.

"We pushed the machines to the bloody limit," Elliott said, "but we kept safety in mind. Freeport has always pushed machines *way* beyond what the manufacturers thought they could do. We have to get a job done, and quick. Because of our location, we can't just wait to get the proper equipment."

One of the Indonesian drillers, Beni Haryono, came up with an original solution to a common problem. Haryono needed to change his drill orientation, and quickly. But the rig is very heavy, and re-orienting it is a cumbersome procedure. The helicopter had just come in, so he talked to pilot Richard Alzetta, one of the steadiest and most capable helicopter pilots working anywhere. "How much lift can you give me? And for how long?" Haryono asked. With the helicopter lifting, and the crew on the rig tugging and pushing, he was able to change his orientation in less than three minutes. This is not a procedure you'll find in the Longyear operator's manual.

Working in Irian Jaya provides some unique challenges for field geologists. Doyle remembers the first time he began geological mapping in the Wabu area. Standard procedure when walking a transect is to measure and mark points with a bit of bright ribbon. When Doyle went back to an area he had walked earlier, he couldn't find his ribbons. The problem is, the Moni just couldn't resist these bright decorations. "Pretty soon," he said, "I noticed that the next fifty people I walked past had my ribbons on their heads, or around their wrists, and my survey stations just weren't there anymore."

The same problem arose with the permanent benchmarks the surveyors used when cutting a grid. They would put an aluminum disk and a steel disk on a survey peg and drive it into the ground. "They'd be coming back the next day and the disk is gone, and somebody's wearing it as an earring."

Drill collar locations suffered the same fate. At first the drillers dropped a piece of PVC pipe into the hole to mark it, but it soon became apparent that this was a big mistake. Plastic pipe makes an excellent container for water, and they were barely in the ground before they disappeared. "I think we've got it beat now," Doyle says. "Now we actually take the rig, burn a rod into the ground, and leave it. And if they can get that out, more power to them."

The exploration geologists are the first people on the ground, and in Irian Jaya, this means walking into areas where the people have almost no contact with the outside world, especially westerners. There is a lot of mutual curiosity, of course, but in a place where some 250 distinct languages are spoken, communication can be a problem. Usually somebody in the group can speak some Indonesian, but even then there are some very large cultural divides to cross to be able to explain why our geologists are walking through streams, putting bits of rock and mud in plastic bags, and tying pink ribbons to the trees. Or, for that matter, running a huge and noisy machine for days just to get trays of rock out of the ground.

After many frustrating misunderstandings, Gary O'Connor has learned to speak in parables. The most successful of these, he says, is the cuscus hunt. The cuscus is a cute little cat-sized marsupial that has the unfortunate luck of being extremely prized for its flesh and soft fur. O'Connor, who has spent lots of time with highlands Irianese, knows how excited the men get on a cuscus hunt. As he puts it, "they just go berserk." The staple in the highlands is sweet potato, and fresh meat is a rarity. If a cuscus or other game is encountered, everyone stops whatever he is doing and tries to catch it.

O'Connor tells them that the geologist's job is just like cuscus hunting. You have to follow "trails" (the transects) occasionally check for "sign" (sampling) and sometimes set "traps" or shoot "arrows" (the drill holes). The Irianese are often puzzled when the geologists come through, mark trails and set up drills, and then suddenly disappear. O'Connor tells them it is just like hunting, you may fire an arrow, but you don't always get the cuscus.

The Eastridge Discovery Zone, an outcrop of arsenopyrite containing high values of gold. In caps are geologists Peter Doyle (in foreground) and Louis Bell.

HIGH ABOVE THE Wabu River there is an old helipad, and just uphill from it is an outcrop of bare, blackish rock. Although the rock is friable and weathered, vegetation won't grow on it. This is the "Eastridge Discovery Zone," the first part of the huge Wabu skarn deposit to have been found by Freeport's geologists. As the skarn began to be mapped in more detail, this discovery zone was found to be its easternmost extent. Wabu is in the Block B exploration zone, about 40 kilometers north of Grasberg.

The vegetation avoids the blackish rock because it contains arsenopyrite, embedded in skarn. "The arsenic is not very concentrated," says geologist Louis Bell, who did much of the early work in this area, "but we try not to lick our fingers when we work with it." This rock does not just contain arsenic and pyrite, a common sulfide of iron, but also gold. It is what geologists call a "high-grader's delight," and grab samples from this outcrop have been assayed at as much as 126 grams of gold per ton. But the Eastridge Discovery Zone is a showing, not a deposit, stresses Peter Doyle. The forces that acted on this rock, turning it from limestone to garnet skarn, came along very late geologically, and it is not typical of the rest of Wabu.

West of the Eastridge, vertical white cliffs stand out from the thick vegetation. This is the only rock marker seen by the layman. Just above the cliffs is the ridgeline. Just below the ridge, Doyle says, is what is called "the marble front," the limit of the influence of contact metamorphism. Or, put another way, the limit of the influence of the hot, high-pressure fluids that are able to turn limestone into marble. At the ridge crest, up over it, and south, a geologist finds only unaltered sediment—calcareous mudstone, siltstone, limestone. North of the marble front, going downhill, he encounters first garnet skarn, at the furthest reach of metamorphism, then magnetite garnet skarn—the white cliffs—then hornfels, a muddy or silty sediment that has been metamorphosed. Beyond the hornfels, which is a poor host for mineralization, are the intrusives themselves.

Currently, the Eastridge is not considered the most promising part of the Wabu skarn deposit. The central part of the deposit, which has been influenced by angled cross-fault structures, has our geologists the most excited. The intrusive itself, although it has a lot of pyrite and other minerals in it, has not yet been thoroughly investigated. It is, Doyle says, "a totally different target."

Wabu has proved to be an important, but frustrating deposit. One can pan gold in all the streams in the area. Geologist Steve Rose came up with one pan sample that indicated 444 grams of gold per ton. Some of the soil samples from the valley test out at 60 grams of gold per ton. That's enough, Doyle says, to almost make the soil itself a resource. But at the same time, the shape and nature of the deposit is elusive. There are very few skarn deposits in the world that are the size of Wabu, two kilometers wide and almost six-and-one-half kilometers long, let alone one that is mineralized. In one place the principal deposit is arsenopyrite, but just 500 meters away it is something else. "This deposit," Doyle says, "is a chameleon." We have brought in skarn experts from outside the company to have a look at Wabu. One of the best, Larry Minart, has visited twice. "Wabu is unique in the world," he says.

One of the big problems is with the deposit's size. It has been very difficult to come up with a detailed structural picture of Wabu, Doyle says, because there are not a lot of "marker horizons" or other identifiable structures that show up in hole after hole. What's obvious is that there is a lot of gold here. What's not obvious is where—and perhaps if—it lies in mineable concentrations. "If I had hundred dollars for every time on this particular project we've come up with an idea that we thought would work, and then went out and put a couple of holes in it only to watch the whole model fall apart, I could retire," Doyle says.

The most significant intercept on the property, which was taken from a drill hole on the "Eagle's Nest," a grassy hill in the central part of the Wabu Valley, totalled 93 meters, and assayed at 8.9 grams per ton of gold. That is a really big intercept. In 1994, we had 160 intercepts at Wabu where the interval was bigger than 3 meters, and the gold content in excess of 1 gram per ton. That is a very large number of gold intercepts.

So far our geologists have identified three parallel zones of low-grade material—under three grams per ton—on the ridge, but we still need some "sweetener" to make the project cost-effective. So the program continues. Geologists continue to take samples, 3,500–4,000 per month, which are analyzed for 12 to 61 elements. So far, 63 sites have been drilled, and 55,000 meters of core samples analyzed. Two drill rigs continue to run through the night, working at 150-meter intervals over the likeliest targets.

Wabu was first identified by Mark Gillian and Gary O'Connor in 1990. They were doing reconnaissance on the whole exploration area, trying to get a sample for about every ten kilometers. They would pick the primary drainages and hit them first. If they got anything interesting, they would follow up. Wabu was detected on the very first pass. There was skarn float in the stream, there was intrusive float in the stream, and gold was showing up in the pans. They sampled the secondary and tertiary drainages, and continued to get interesting results, so Wabu became a prospect.

Mapping and groundwork continued, and in 1992 the ground program started taking samples every five square kilometers. Things still looked good, so the base camp moved in early 1992 from Karubaga, a village in the Dani area 100 kilometers east of Wabu, to Bilogai, near the prospect. Hoisting began in the district, and then geologists started mapping and sampling work along drainages near the camp. In 1992, Freeport flew a helimag survey of the area, and Wabu really stood out. In fact, the anomalies define the central zone of the deposit. There is one point on the aeromag printout that looks exactly like a bull's eye. This has been named the Apex Anomaly, and

Breaking down rods on a rig in the Wabu area. To date, more than 55 kilometers of core have been extracted and analyzed, and drilling continues.

lies right on top the ridge. Later, when we put a drill hole in it, we intersected 100 meters of magnetite skarn, which is a very significant magnetic anomaly.

In October 1992, the work at Wabu got a boost from our CEO himself. The geology team at Wabu met with Moffett in Tembagapura, and he was very impressed with the sampling results. Although our geologists were still being stymied by the sheer size and complex shape of the Wabu deposit, he suggested we start a drill program. Rigs were brought in immediately, and started work started on targets spaced at half-kilometer intervals. The result of this high-speed, 12-month program was—well, confirmation that exploration programs are not easy. Indications of gold mineralization were coming in, but as Doyle says, "we didn't hit a Grasberg on the first hole."

The geologists went back to the basics, reworking the promising areas and collecting more samples. They began to notice that the hits came in areas where there were deeply cut creeks crossing the ridge, in other words, cross features. Then they started to hit more consistently. We in New Orleans became interested, and still are. Now the

This rig working a hillside in the Etna Bay district has been struggling with loose rock, and the crew is pulling the rods to replace the diamond bit.

drillers are working 25-meter step-ups in both directions from the best targets.

When Moffett gave the go-ahead in October 1992 to speed up the program, the base camp at Bilogai changed overnight. Additional bunk houses sprang up, as did two or three supply sheds and a core logging facility. At the same time, the geologists were out walking transects and construction crews were building drill pads all over the valley and on the slopes of the Wabu Ridge.

"It was a very, very hectic time," Doyle says. "*Every* piece of material that came in for that expansion came by helicopter. Some days it looked like something out of *Apocalypse Now*. You'd get four helicopters in a row coming over the hill with drill parts, rods, or fuel. Everything was coming over the top. It was phenomenal. I've never seen a program take off like that."

This project ended up costing more than $20 million, and we still don't have what we consider to be sufficient critical mass to justify a project. But this is the nature of exploration. Wabu was our first major prospect outside of the immediate area of the old Ertsberg and Grasberg areas, and we learned a lot. The logistics people, the drillers, the geologists—everybody involved with our exploration program anywhere in Irian Jaya—was trained at Wabu. We now have a core of trained, capable people who are able to function at high efficiency and top speed. As a result, for example, the Etna Bay startup in 1994 was quick and relatively painless.

It is important to remember that enough gold has been found at Wabu that, if it were anywhere else in the world, operations would already have started. But our costs in Irian Jaya are so high, and logistics so difficult, that we require a really rich, well-defined resource to start laying in the infrastructure necessary to mine it. Although we have confidence in the Wabu area, it will take time before we are sure. In the meantime, there is plenty in the area to keep our drillers and geologists busy. And in addition to Wabu, we are looking at Mendoga, Hitalipa, Yabu and Minjauh, sites named for the nearest village, and Kemabu, named for a nearby river. Crews are busy in these areas with preliminary ground reconnaissance, exploratory drilling, traversing, ground magnetics, sampling and geological mapping.

In August 1994, Freeport formed a 90 percent ownership subsidiary called P.T. IRJA Eastern Minerals Corporation—we call it, informally, Eastern Mining—and signed with the Indonesian government to explore and develop resources in three tracts of land, totalling 2.5 million acres. The biggest tract, and the most promising to date, has been the 10,000-square-kilometer Block I area inland of Etna Bay. At first glance, this looks like the last place in the world to look for minerals. It includes many square kilometers of low, swampy land, and the mountains here stand outside the main highlands belt. The broad swamplands around the Omba River, east of Etna Bay, have such a prehistoric appearance from the air that our pilots and geologists have taken to calling the area "Jurassic Park."

THE ETNA BAY Base Camp sits on the water's edge, at the innermost reach of Etna Bay, a narrow, very well protected, deep water bay. The outer reach of the bay is called Lakahia Bay; the innermost takes its name from the Dutch ship *Aetna*, itself named for the famous mountain in Sicily. Personnel and equipment arrive at Etna either by barge or by seaplane, a rugged little Grumman Mallard that makes an almost daily run from Timika. The camp is also just a short walk west of Etna village, or Kiruru, a source of labor and the administrative seat of the eponymous subdistrict.

Eastern Mining has built an efficient and comfortable little village out of wood, tents and pre-fabricated materials, with warehouses, medical facilities, a communications room, a work space for geologists, and living and dining facilities. All the buildings are raised a meter or two above the damp ground, and are connected to each other by sturdy plank walkways. Anywhere from 15 to 20 permanent employees work at the base camp, providing logistical support for the 100–150 geologists, drillers, and local crew members working in the district. The base camp, with its satellite television good showers and hot meals, provides a nice decompression stop for geologists working out in the fly camps.

From the helicopter pad, it is a very short flight over the 900–1,500-meter coastal mountains to the south and east, where exploration and drilling activities are concentrated, although morning fog and afternoon rain often keep the choppers on the ground.

Etna Bay marks an abrupt change in Irian Jaya's geography and geology. East of the bay, the wide alluvial plain and silty rivers and bays that characterize Irian Jaya's south coast begin. West of the bay, limestone cliffs plunge straight into the sea and the bays are sparkling clear and filled with coral. Just 15,000 years ago, New

The base camp at Etna Bay. This comfortable little village is the center of Block I exploration activities, and supplies the scattered fly camps and drill camps.

Guinea and Australia were connected by land. The border of this land bridge started at what is now Etna Bay's eastern shore.

Exploration work around Etna started in late 1993 under a pre–contract-of-work agreement, and was largely inspired by a few bits of information uncovered by Gary O'Connor, now the project's chief geologist, during a literature search. "Originally when we were applying for ground, we were looking at the northern metamorphic belt. I'd done some literature research—it was an old Dutch report, a compilation put together in 1960—and I'd noticed a reference in it to intrusives, mineralization, and—they were actually oil geologists—to what they called 'ore' in the Etna Bay area. The reference was just a little aside, the whole report was put together for oil."

So O'Connor made a proposal. The stratigraphic sequence is the same as what we had in the highlands, the geologic setting seemed generally similar, and the intrusives in the Etna area were of about the same age. And some oil geologists had noticed "ore." So we flew an aeromagnetic survey in early 1994, and as these results also looked promising, we acquired the ground in August 1994.

Freeport's Grumman Mallard lands on Etna Bay. The running cost of this little plane is less than that of a helicopter, and it can fly faster and in worse weather.

Field geologists working the Etna prospect face a physical and cultural setting very different from Wabu or the highlands. Where Wabu is comfortable, even chilly at night, and scattered with Moni villages, Etna is hot and steaming, and almost unpopulated. Etna is a dense, lowland rain forest, rich in plant and animal life, but most of the people here live along the coast. The climate is such that, as field geologist Philip Petrie says, "When I've got those gum rubber boots on, I have to empty them twice a day."

Petrie works out of a simple fly camp in the Timini area, walking the streambeds to refine the geological map of the area. The camp is a few tents and a cookstove at the edge of a clearing. A helicopter pad stands in the middle of the clearing. A small stream serves as a shower and water tap. The camp is basic enough to be broken down, packed up, and hauled to the chopper pad in less than an hour.

Petrie is a short, solid man who wears what he calls "the Aussie uniform"—a pair of shorts and a T-shirt—under his yellow field vest. There aren't too many leeches in the area, but when your legs are bare it's easier to spot them. The creeks Petrie is checking are anywhere from fifteen minutes to an hour's walk from the fly camp. The walking, under the rain forest canopy, is fairly easy, as the trees so block the light that the undergrowth is sparse.

At the lower end of the streams, the bedrock is fine mudstone or siltstone, and as you work your way up the stream, blue stains begin to appear on the brownish rock. These are from calcium in the water, and are an indication that there is limestone upstream. The calcium deposits make the rock less slippery, which is welcome. Because an ore deposit usually requires both an intrusion and a good host rock, Petrie says, there are two ways to go about looking for mineable deposits. "You can look for intrusives, or go out and look for the right sediment type."

Petrie, and the other geologists working this area, pay careful attention to any outcrops of bedrock along the streambed. The directionality of the fracture lines in these outcrops are carefully noted, and entered into a larger map. It may be the case that this fracturing was caused by the forces of a nearby intrusion, and Petrie is hoping that the pattern of the fractures, when recorded from all the creeks, will point to an intrusion somewhere. This is particularly important in the area Petrie is working, as from aeromag and other data, it looks like the nearest intrusion lies beneath a wide, crocodile-filled swamp, which rather hampers direct access to the rock there.

As he works his way up the streams, Petrie chips away some of the bedrock and tests it with a small bottle of hydrochloric acid. Eventually, when he reaches the limestone, his samples will bubble furiously. The course of the stream, the analysis of the rock type, and the presence and orientation of the fractures will, at by the end of the day, be preserved as a neatly written page in his waterproof field notebook. There is an optimal number of readings, Petrie says. Some of the newly trained Indonesians, in their enthusiasm, would take readings every twenty-five meters, but they soon learn to slow down. Data from all the field geologists, transferred to a 1:5,000 scale geology map, will shape the hypothesis that directs the drilling and exploration work here.

Although accommodations at the fly camp are spartan, Petrie, who worked for a number of years in Papua New Guinea on the other side of the border, says they are almost luxurious compared to his days on the PNG side. His team consists of himself, two assistants, and a cook. Because the camp has been established for a while, the sleeping tents are raised off the ground on wooden platforms, something of an added luxury.

Some of the geologists have called our current program "the Rolls-Royce of exploration." Petrie's exploration teams in Papua New Guinea consisted of two geologists and food and equipment for ten days, and everything had to fit comfortably into the payload of a Bell 206 helicopter. Freeport uses Bell 205s with triple the payload of the 206, and our fly camps include assistants and cooks.

Some of the geologists are uncomfortable with the speed at which we are proceeding with drilling and some of the later stages of exploration. "You'd still be walking up a creek when they're landing a drill rig on you," says one. Drilling is taking place in the Etna area even before the area is well-mapped geologically. O'Connor says some members of his team find the pace a bit fast, which they feel compromises their ability to do a thorough, professional job. But, as he puts it, "the holes are going to have to be drilled eventually. And some companies, their problem is the reverse. They'll chew things to death, just scratch and poke, and waste a lot of time before putting the drill on. Freeport doesn't do that."

One of the reasons we bring in drill rigs so early is that it is so difficult to obtain accurate surface geological data in Irian Jaya. Where the geology is not obscured by vegetation, rapid tropical weathering makes it impossible to determine the mineralogy of the surface rock. In our initial drilling we are often not aiming at an ore target, but simply trying to acquire basic geological and structural information about the underlying rock. The extremely high logistics costs of running an exploration program in Irian Jaya also encourages us to proceed as rapidly as possible.

T HE DRILL CAMP in the Etna prospect is just a few minutes from base camp by helicopter. Twenty-eight men work here, on two rigs, all under the leadership of Pardal Suyatno, the 41-year-old Javanese supervisor. Suyatno has been a driller for 15 years, and he likes working in Irian just fine. The food in the mess is good, the camp is comfortable, and the rock no more or less uncooperative than anywhere else he has worked.

In early 1995, two rigs were working in the vicinity. One was quite near the crest of the peak, and the drilling there was easy. The other worked the side of the mountain, driving its bit at an angle. Although only a couple of hundred meters from the drill camp, it was a couple of hundred meters straight down. Access to the drill rig was provided by a custom-built wooden staircase laid into the side of the ridge. By early May of 1995, the crew had already spent three weeks reaching 600 meters. The geologists wanted 900 meters, and Suyatno figured that might take another three weeks. This was a very stubborn hole, as the rock was broken and loose, and the diamond bit would just spin futilely, using all of its energy to wear itself out. The word Suyatno used for this particular rock was *jelék*, literally "ugly."

Although it may seem counterintuitive, hard, dense rock is far easier to drill than loose, broken rock. The latter is the bane of drillers. Sometimes they can pour polymer additives into the water lubricating the bit, and this will help stabilize the rock and "heal" the fractures. But usually there is nothing to do except be patient, and keep trying new bits. The drilling rate averages about 30 meters a day at Etna, and about 35–40 meters a day at Wabu, where the rock is generally more competent. If they can meet or beat this rate on a given hole, the drill foreman can enjoy a bonus of sometimes $1,000 per hole, which is quite an encouragement.

When it comes to equipment, drillers are a conservative lot. Most of the older ones swear by the Longyear 44, a heavy-duty machine that is about as reliable as a block of granodiorite. But heavy-duty also means *heavy*, and one of the revolutionary aspects of our drilling program is our early reliance on the Longyear LF70, a light, modular, hydraulically driven rig. The head of our drill program, Fred Elliott, was the first to bring these new rigs to Asia. He wears a tie now, but he has plenty of field experience, including three years drilling for us in Irian Jaya.

In reading one of the industry journals, Elliott came across a story about a "funny little drill rig" designed by a man named John Stevens in Coeur d'Alene, Idaho. Elliott called him up. Apparently John Stevens, who is a fascinating jack-of-all-trades—engineer, welder, hydraulics expert, and machinist—was working in Alaska, where he was moving Longyear 44s around by helicopter. He thought to himself that there had to be a better way, and actually designed a new style of unit in his tent at night. The company he was working for told him to go ahead and build it. Around 1991 and 1992, when we started getting ready to drill on Wabu, Elliott took a gamble on these new rigs, and had two sent to Irian.

Lugging a heavy core tray up a near-vertical plank walkway to the chopper pad. In Etna, an average of 30 meters a day from each hole are flown off for analysis.

The LF70 rig uses a hydrostatic drive and is light and easy to disassemble. Where a Longyear 44 takes 15–20 helicopter loads to transport, this unit takes a maximum of seven. With helicopter time running about $2,000 an hour, it took no more than three months for the rigs to pay for themselves in savings on chopper time alone. And the depth capacity and reliability have been just as advertised.

"Freeport does things differently," Elliot says. "Back in the late '80s, we were doing a lot of underground drilling out at the mine, and I settled on a new type of hydrostatic machine for underground work. And I was told by people in senior management of the Longyear Company, 'This is going to be failure because Indonesians will never be able to operate one of these things, they'll never be able to understand it, it's state-of-the-art, blah, blah blah.' I said: 'Watch me.' We went through a severe learning curve, but were incredibly successful.

"The Freeport difference is, if someone has a better idea or has a different idea from what is being done in the rest of the world, and is willing to put his heart into it and make it work, management will back you and you just do it. And if you fall on your face, well, you fall on your face. You never learn unless you screw up sometimes, right?"

CHAPTER NINE

The Use and Sale of Copper

An ancient metal—The Cyprus copper works—Copper and electricity—The flotation revolution—Smelting and refining—Heap leaching—A recyclable metal—The custom market—World demand

COPPER, LIKE GOLD and silver, occasionally occurs in nature in a native, elemental state—that is, as a nearly pure metal. For this reason copper became one of the earliest metals to be used by mankind. Archeological evidence suggests that copper first came into use more than 10,000 years ago in an area that includes what is now the eastern part of Turkey and part of Iran. This period, coming at the end of the Stone Age, has been called the Chalcolithic ("copper-stone") Period.

Pieces of native copper, gold and silver were shaped into useful objects by striking them with stone implements. It was 4,000 years before smelting technology developed, and the earliest known smelter was operated about 5,800 years ago at a site in present-day Iran. Pure copper, of course, is quite soft, and bronze came into use soon after the introduction of smelting. This first alloy probably resulted from the accidental introduction of arsenic or antimony into the melt, and later tin. Thus began the Bronze Age, in which technologies took advantage of the superior strength of copper alloys.

In North and South America, the use of copper extends back perhaps 3,000 years, beginning with various American Indian groups in Central America and the Andes. These cultures never developed bronze or brass, and smelting methods were crude and on a small scale. These societies' use of copper was not intense. On the island of New Guinea, copper metal was unknown until it was introduced by traders from the western part of the Indonesian archipelago and Europeans. Native copper occurs only rarely in New Guinea, and the copper-bearing deposits in Irian Jaya lie in the remote and barren highlands, where the Irianese rarely ventured.

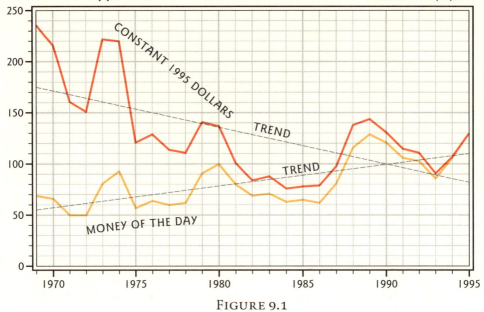

FIGURE 9.1

The development of bronze technology is also associated with the rise of some of the early great civilizations in the Near East, such as Mesopotamia and Sumer. The scarcity of alloy metals, particularly tin—which became the most important additive metal about 5,000 years ago—prompted a search for raw materials in new lands. Many historians think this search stimulated the expansion of ancient Middle Eastern civilization into Europe and Africa, and possibly Asia.* With this search came the development of commerce and specialized labor, and technologies for mining, smelting, and casting. The Romans further refined copper and bronze technologies, and these were carried through the Middle Ages to the present day.

The word "copper" comes to us from the Greek *kuprios*, from *Kupros*—the island of Cyprus—the origin of much of Greek and Roman copper. Cyprus was blessed with rich deposits of malachite and azurite, minerals containing more than 50 percent copper, and

---

*Some experts believe that copper smelting and bronze technologies developed independently in Asia.

248

Dried concentrate being loaded in a ship at Amamapare. Freeport's concentrate is exceptionally 'clean'—free of impurities—and is in demand by smelters.

over the millennia, resins from the island's famous pine groves mixed with the groundwater and naturally reduced the ore to native copper. By 2760 B.C. the Cypriots were smelting their rich ores by burning them with umber, an oxide of manganese and iron that is the source of the familiar brown and red-brown pigments. Because Cypriot umber—more than 50 percent iron—was used in copper smelting, some historians suggest that the Iron Age began on Cyprus.

Cyprus became a much more important source of copper and bronze than earlier operations in the Middle East, which quickly ran out of wood to use as fuel in smelting. It has been estimated that the entire forest cover of the island had to have been replaced 16 times over the productive life of the Cypriot mines to provide enough charcoal for the island's smelters. Many roads in Cyprus still rest on a bed of centuries-old slag. Before the island's mines and smelting operations ran out of ore and wood, Cyprus produced some 200,000 tons of copper and bronze. For reference, this is about five months production from Freeport's Grasberg mine or, put differently, it would supply the world at its current rate of use for one week.

249

Of all the metals, copper has been second only to iron in its use by mankind through the ages, and it retains this status to the present day. The modern history of copper begins with the discovery of electricity, and the wide range of inventions that followed. Today, in parts of the world where standards of living are high, so is the consumption of copper. Worldwide, the predominant use of copper is in electrical equipment and supplies, and in highly developed economies such as Europe, North America, and Japan, electrical applications comprise approximately 60 percent of the use of copper.

Copper and its alloys have many properties that make them useful in a wide range of applications and products for the modern world. In antiquity, it was the metal's good workability, the ease with which it could be alloyed, its hardness, its resistance to corrosion and its attractive color that made copper useful. These properties remain important today, but some of the metal's other characteristics have assumed greater importance. These include copper's combination of high electrical and thermal conductivity and high strength, and the ease with which it can be welded and soldered. Its corrosion resistance, particularly in modern alloys, provides many applications in the manufacture of industrial equipment, especially parts exposed to acids, and in the marine and aircraft industries.

Copper is widely used in the telecommunications industries, but recently has lost some of its growth to fiber-optic cable. However, even though optical fiber is replacing copper as cabling, the connectors and switching units still rely on copper, and the efficiency and high bandwidth of this new technology has created a new demand for copper in the manufacture of devices to connect to both ends of the fiber-optic cable: computers, telephones, and facsimile machines.

Copper has a myriad of non-electrical applications, including industrial machinery and parts, household and commercial air conditioning equipment, farm machinery, roofing, downspouts, gutters, nails, rivets, soldering materials, seals, heating and cooling and water carrying tubing, radiators, brake pads and linings, bushings and bearings, and carburetor parts.

One of the best known uses of copper is in the manufacture of ammunition. Brass, an alloy of copper and zinc, is used for cartridges and shell casings. Sometimes, copper is called the "red metal" because of its color. Sometimes it is called the red metal because it is so heavily used during wartime.

IN PAST CENTURIES the standard practice of the mining industry was to move copper ores directly from the mine to the smelter, with little if any effort to first concentrate the copper minerals. But with the increasing use of electricity in the first decade of the 20th century, and the sharply increased demand for copper that this produced, this situation began to change. The problem was ore grade. The accessible, high-grade deposits of antiquity were for the most part played out, and the new demand for copper required that lower grade ores be mined. These ores require separating the copper minerals from the waste, or "gangue," so they can be economically shipped to the smelter. Early concentration techniques relied on hand sorting and gravity methods, but these proved inadequate.

Then, in the first decade of the 1900s, a discovery took place which changed not just the copper industry, but the entire base metal industry. This was the invention of froth flotation, an efficient and economical method of concentration that is used in most base metal mine operations today. This invention is so important that it is safe to say that if it weren't for froth flotation, the modern civilization and world economy we see today could not exist.

In froth flotation, ore is crushed and ground to a fine particle size, mixed with water into a slurry, and then, after the introduction of small amounts of reagents and air, passed into a chamber with vigorous aeration. The desired ore minerals are air avid, and thus attach themselves to the bubbles of air, forming a foam on the top of the chamber that can be easily skimmed off. The gangue is water avid, and remains in the chamber where it can be rejected as waste. Froth flotation enables the industry to mine low-grade ores of many different types, and to concentrate the metals into a form that can be economically shipped and refined. (See also Chapter 7, page 207.)

Concentrates of the highest possible grade usually require several stages of flotation, each stage increasing the purity of the product. The copper content of a given concentrate depends chiefly on the mineralogy of the ore, and to a lesser extent on the thoroughness and efficiency of the flotation process itself. Typical copper concentrate produced today contains between 25 percent and 35 percent copper, with very rich concentrates sometimes containing as much as 50 percent copper. Our concentrates contain 29–31 percent copper, and additionally, .7–.9 troy ounces of gold and 2–3 troy ounces of silver per dry ton.

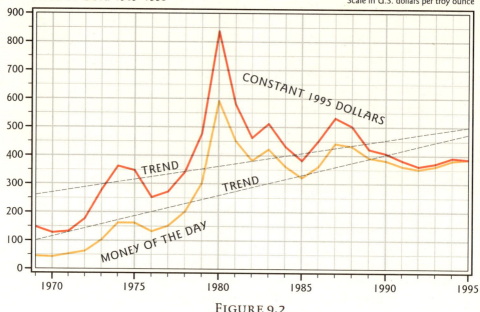

FIGURE 9.2

The copper smelters of the world, who are our customers, generally consider Grasberg concentrate to be ideal in composition and extremely clean. It contains the proper balance of sulfur (35 percent) and iron (25 percent), a combination which, because it is combustible, helps maintain the heat of the furnace and produces a desirable slag. The concentrate is considered clean because it contains extremely low quantities of undesirable elements, such as antimony, arsenic, bismuth, fluorine, and zinc. Freeport concentrate sets the standard for the industry, and it is eagerly sought after for blending with concentrates of lesser quality. The addition of some Grasberg concentrate can make a poor quality concentrate workable, or help a smelter meet its pollution-control requirements. Many concentrates require mixing, and Grasberg concentrate is one of the few that a smelter can economically use for 100 percent of its feed.

Today, about 85 percent of all primary copper—that is, copper metal produced from mining rather than recovered from scrap—comes from sulfide concentrates. (Sulfide ores, like Grasberg ore, are far more common than other types, such as copper oxides.) These

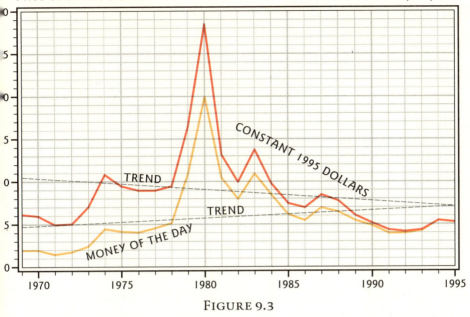

FIGURE 9.3

sulfide concentrates are processed exclusively by pyrometallurgical methods, that is, by smelting. Smelters are very expensive to build and run, and are generally only economical for mines producing over 100,000 tons of copper a year for an extended period of time. The high capital cost of building a smelter, combined with the relatively low cost of transporting concentrated ore, has led to the construction of custom smelters that carry out processing for mines whose limited output or modest reserve size do not justify a smelting facility of their own.

In the early years of the Freeport mine, when only Ertsberg was being mined, the limited size of the deposit and scale of operations did not allow us to consider a smelter devoted to our ore. But with the increasing scale of operations and our greatly enlarged ore reserves, this situation has changed. We first purchased smelter capacity from Rio Tinto Minera (RTM), near Huelva, Spain, an operation whose mine was nearing the end of its useful life. The RTM smelter underwent an expansion to accommodate the Freeport ore. Freeport has also joined with Mitsubishi to build and operate a copper smelter at

Gresik, near Surabaya in East Java. At the time of this writing, we are in the final stages of a feasibility study for the Gresik smelter, and initial production from this facility is expected in late 1998 or 1999. By this time, the combination of the RTM capacity and the Gresik smelter could account for 70 percent of Grasberg's output.

The key to many copper smelting operations are customers to purchase sulfuric acid, a byproduct of the smelting process, and customers for the refined metal. It is cheaper to ship copper concentrate in bulk form to a smelter located near sulfuric acid and copper metal buyers than to smelt near the mine and ship the sulfuric acid and refined metal. The smelters in both Spain and Gresik are near sulfuric acid and metal customers.

Although there are many variations in smelting methods, the basic pyrometallurgical process is the same for all the smelters of the world. First the concentrate is smelted, basically burned, using pure oxygen and perhaps the addition of fuel, depending on the composition of the concentrate. Fluxes, such as silica, are added to facilitate the formation of slag, to carry away the impurities. The initial enriched copper sulfide, called "matte," flows out of the combustion chamber in one stream, the slag in another. The matte, at this point perhaps 75 percent copper, is delivered to the converter, where it is blown with pure oxygen until additional slag is formed, as well as "blister" copper. Blister copper, 96–99 percent pure, is subjected to a cleaning step to remove excess oxygen, and then cast as slabs of nearly pure copper.

These slabs are called anodes, as they will serve as the anodes in an electro-refining process. The anodes are bathed in a vat containing an electrolyte of sulfuric acid across which an electric current is applied. The anode dissolves in the electrolyte and is re-deposited on the cathode side of the cell. In this process the impurities, which include gold and silver, form a sludge on the bottom of the tank as the anode dissolves. The cathode copper that results is extremely pure, typically better than 99.99 percent copper. It is important to keep the purity of the product very high, as the presence of impurities in the copper can affect its electrical conductivity. Standards exist in the copper marketplace to assure the customer a quality product.

Smelting is not the only way to recover copper from ore. Oxide ores, and a few sulfide ores with favorable mineralogy, at times can be processed by hydrometallurgical methods, generally called heap leaching. This type of processing, which yields a high-quality copper metal product, has taken a qualitative and quantitative leap forward since the 1970s. Leaching involves placing a "heap" of crushed, unconcentrated ore on a specially prepared surface or "pad," and soaking the ore heap in a solution of sulfuric acid. The acid, in the presence of air, leaches copper from the ore, and the resulting enriched solution is collected from the pad. Often, waste and overburden dumps contain enough copper that they can also be treated in this way.

The most common leaching method in use today is called solvent-extraction–electrowinning, or SX–EW. In this process, the copper-bearing solution of sulfuric acid that drains from the heap is collected and concentrated and "electrowon." The electrowinning part of the process is very similar to the electro-refining process described above, and yields a high-quality copper that competes with cathode copper in purity. Where the mineralogy of the ore and conditions of terrain, rainfall, and climate suit, SX–EW allows the treatment and extraction of metals from low grade ores that could not be handled economically by any other method.

Western world copper production by this means increased from 650,000 tons in 1991 to 1,000,000 tons in 1995—a 53 percent jump in just four years. The use of this process will continue to grow. However, there are limits to the amount of ore that can be treated this way, and even ores that are amenable to heap leaching are often underlain by sulfide ore, which cannot be thus treated. Today perhaps 15 percent of primary refined copper comes from SX–EW, and although there is still room for growth, geology, chemistry and economics impose an upper limit of probably 30 percent on the amount of copper that can eventually be produced this way.

With the high stripping ratio at Grasberg—3.7 tons of waste must be removed for each ton of ore—our engineers have looking longingly at heap leaching as a means to produce value from the overburden that we must move to reach the ore. We have conducted numerous sampling programs, experiments, and studies to evaluate the potential. To date, with the conditions and ore composition at Grasberg, we cannot see a way for the process to do better than, at best, break even. With further developments in the industry, such as bacterial leaching and maybe higher sustained copper prices, perhaps the situation we see today will change.

Copper can be melted down and reused indefinitely, and relatively

Copper anodes being cast at the smelter. Under the influence of heat and oxygen, the impurities in the copper concentrate form slag, leaving nearly pure copper.

inexpensively, and it is impossible to tell copper that has been recovered from a former use from primary or newly mined copper. As a result, each year approximately 37 percent of the total production of copper produced as copper metal or as a copper alloy has actually been reclaimed from scrap. At first glance, this seems to be a low rate for such a valuable and relatively easily recyclable material. This statistic merits further investigation.

Copper is a very durable commodity. Once it is placed in use for whatever purpose, say electrical wiring in a house, or industrial machinery, it will remain in use for years. The time in use will vary. For example, an automobile may last 6–8 years, while a building may last 60–80 years or longer. Some experts have estimated the average copper product life at 33 years. Because the use of copper continues to grow, for a true calculation of the recycling rate, one must take into account the product life. In 1962, the world consumed 5.7 million tons of copper metal. If 33 years is a reliable estimate of copper's product life, the 4.45 million tons of copper being recycled this year actually represents a 78 percent rate of reuse.

These copper anodes, already 99 percent pure, will be placed in an electrolyte and, under an electric current, redeposited as 99.99 percent pure cathode copper.

The high percentage of recovery and reuse of copper, therefore, designates it as an environmentally friendly material. Energy, labor, and materials are conserved from the mining concentration and smelting process. Landfill problems associated with disposal are reduced. And recycling extends the world's mineral resources.

UNTIL VERY RECENTLY, all of Freeport's sales of its only product, copper concentrate, were made on what the industry calls the "custom concentrate market." The custom market consists of sales by mines of their concentrate to smelters that do not have integrated mine facilities. For the most part, custom smelters are located near ocean ports where very large vessels, up to 35,000 tons, can deliver their copper concentrate.

Approximately 40 percent—2.8 million tons in 1995—of the western world's newly mined copper production is supplied to custom smelters by mines with no direct association with those smelters. This 2.8 million tons of refined copper represents a total of about 8.4 million tons of copper concentrate. In 1995, Freeport produced approximately 500,000 tons per year of copper, which represents approximately 1,600,000 tons of concentrate. Some 450,000 tons of concentrate were sold to our affiliated smelter in Spain, leaving 1,150,000 tons to be sold on the international concentrate market. This amount represented 13.6 percent of the entire world custom market in 1995. In volume, this sale of concentrate was surpassed only by Chile's Escondida (17 percent of the custom market), and P.T. Freeport Indonesia was followed by Ok Tedi (7.6 percent) in Papua New Guinea, and Highland Valley Copper (5.8 percent) in Canada. Once the smelter in Gresik becomes operational in 1998 or 1999, it will process a large portion of the Grasberg concentrate and Freeport's share of the custom market will drop to 5 percent.

The majority of the custom smelters are located around the Pacific Rim, in Japan, Korea, China, the Philippines, and Australia. Japan dominates this market, and Japanese smelters buy approximately 45 percent of the total concentrate sold on the international market. As Freeport's production increased, it became necessary for us to expand our sales out of the Pacific Rim and into the Atlantic Basin. Our association with the Rio Tinto Minera smelter in Spain provided the opportunity to enter the very important European market for refined copper.

Copper concentrate is sold under two general types of contracts, "long-term" contracts, which by our definition are two years or longer, or "spot" contracts, which usually cover a single shipment of concentrate, or are in effect for less than two years. Long-term contracts, usually signed directly with smelters, form Freeport Indonesia's base load of contract obligations. A smaller percentage, 25 percent or less, are with a group called "traders."

As their name suggests, traders are commercial entities who buy concentrate wherever it may be available, and sell and ship the concentrate to a smelter, hopefully, of course, clearing a profit in the transaction. Through the years, traders have performed an important task for Freeport Indonesia. They have facilitated the sale of our concentrate to smelters who, for one reason or another, were not prepared to give us a long-term contract. Traders also were able to cover the risk of shipments of concentrate to tricky or unstable markets such as China, Russia and Africa.

A commercial transaction for the sale of concentrate is divided into two parts. The sale of the contained metal itself is very straight-

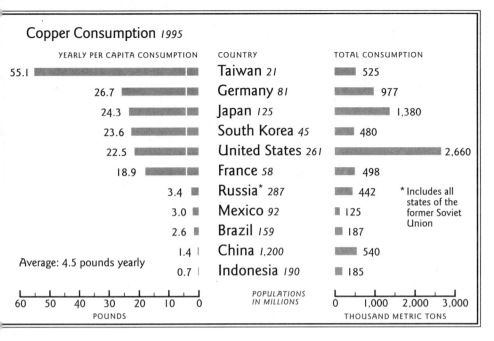

FIGURE 9.4

forward. Freeport is paid for the metal contained in the concentrate at the prevailing monthly average metal price set by the London Metal Exchange. The month that will determine the price, usually one to two months after the concentrate leaves the dock at Amamapare, is agreed upon ahead of time, and the price paid is usually the average for the agreed-upon month, rather than the price on some single day. After the concentrate is actually smelted and refined, an adjustment is made for smelter losses.

The second part of the transaction involves a deduction for the cost of smelting and refining of the metal. This is known in the industry as TC/RC, for treatment charges/refining charges, and is negotiated. The treatment and refining charges can be regarded as a commodity, because they vary according to all the rules of supply and demand. If there is a shortage of smelting and refining capacity in the world, mines are forced to bid up and pay more for this service. Conversely, if there is an excess of smelter capacity, smelters compete for scarce copper concentrate and the mines enjoy lower treatment and refining charges. Through the years, TC/RC costs have fluctuated

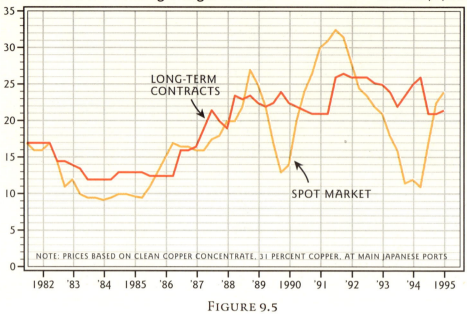

FIGURE 9.5

sharply, but overall have favored the miners. Changes in spot-market contract terms lead those in the long-term market, as the spot market is more responsive to the supply/demand situation at any given point in time. (See FIGURE 9.5, above.)

A rise in copper consumption has accompanied the overall growth of the world's economy, and copper use is generally favored by increases in the standard of living. In general, a given nation's standard of living is mirrored by the intensity of its use of copper. The world's population consumes an annual average of 4.5 pounds per person. However, in countries where the standard of living is high, such as the United States, Japan, Germany, and France, copper usage is four to five times this amount. (See FIGURE 9.4, page 259.) Developing countries, where the majority of the world's population resides, today use one-third or less than the world's average. This is a sharp contrast, but it presents an opportunity for the world's copper industry. In the coming decades, most people believe countries such as China, India and Indonesia will improve their standards of living, and thereby require an increased supply of copper.

In fact, this process is already taking place. From 1970 through 1995, the world demand for copper increased an average of 1.9 percent annually, now requiring more than 10 million tons of copper per year. This statistic, however, masks the growth that is taking place in the developing world. For example, copper consumption in North America and Europe has increased an average of approximately 1.5 percent a year over the last 25 years, while consumption in the developing world over the last five years has advanced more than 7 percent yearly, although from a much smaller base. This trend continues today, and although there are occasional spurts of consumption in the developed world, the real growth in copper demand is coming from the developing world, which now consumes 2 million tons of copper a year and continues to need more.

The world today uses 10 million metric tons of copper a year. If growth continues at today's levels, 1.7 percent annually, the world will need another Grasberg deposit every three years. And this calculation does not account for the decreasing grade that mines will face as they work their way through their deposits, or the closure of old mines due to the exhaustion of the ore bodies. Although there have been short periods of oversupply, worldwide demand for mined copper has generally exceeded supply over the last ten years. Projecting this trend ten years into the future, taking into account deposit mine-outs and lowering head grades of ore, the situation looks quite good for the mining industry.

According to 1991 statistics, the world reserves of copper stand at 321 million metric tons. At today's rate of consumption, this is less than 30 years of known or measured reserves. However, with improvements in mine and ore processing technology, and the additional exploration encouraged by higher metal prices as the supply shrinks, more copper should become available.

This impending shortage bodes well for the copper industry. When shortages exist, we see higher copper prices. When there is excess copper, we see lower prices. With rising demand and a shortfall in production, we can expect low stocks and higher prices.

In real terms, however, it may be that copper is becoming more of a bargain for the consumer. While the apparent metal price as seen in unadjusted "money of the day" has risen over time to where today the market price of copper is well over a dollar a pound, in real terms copper is cheaper than it has ever been. When the prices are converted to constant dollars, since at least 1969 one can see a steady decrease in the real price of the metal. From this standpoint, it is the final consumer who wins. Also, perhaps the lowering of the real cost to the consumer is one of the major factors for the increase in consumption, and by association, the higher living standards of today's consumers. (See FIGURE 9.1, page 248.)

Since Grasberg is a copper mine with an important by-product—gold—we should also consider what has happened to the price of gold. In both real and adjusted dollars, this commodity has gained steadily in value since the early 1970s, when the Ertsberg mine began. (See FIGURE 9.2, page 253.) In stark contrast to copper, the trend for the constant dollar price of gold points to a doubling over the last 25 years. If this trend continues, it is conceivable that Grasberg will one day be first a gold mine and only secondarily a copper mine.

PART IV
Resources and the Future

Making a brick from tailings. The large tailings deposition area will provide a resource for farming, and perhaps eventually, further mineral extraction.

CHAPTER TEN

Managing the Environment

Muddy rivers—Sheet flow—Tailings deposition—Overburden—The shrinking glaciers—A biologically rich setting—hydromulching—Farming the tailings—Recycling—Environmental monitoring

A VISITOR GLANCING out his window as the jet banks for landing at Timika airport encounters an interesting sight: the wide, corded Aikwa river runs ash-gray. The south coast of Irian Jaya is not a land of clear, fast-flowing rivers; all are wide, meandering and heavy with sediment. But in the area, only the Aikwa is gray. The rest are chocolate brown, the color of New Orleans café au lait—or the Mississippi River. The Aikwa's distinctive color comes from the unique nature of its sediment load, a large percent of which is made up of "tailings," the rock flour or silt that is the waste product of Freeport's mill and concentration plant. The copper-rich concentrate piped down to the coast represents only about three percent of the rock that moves through the mill. Everything else is discharged into the Aikwa river system, currently 120,000 metric tons a day.

Physically, there is little difference between the tailings and natural sediments in the river. The striking color difference comes from the particular type of rock, and the fact that the tailings are "young" sediments. Having been ground up and produced by the hand of man, they were not long exposed to air, which would have oxidized their natural iron content, producing the brown color seen in other streams. The issue for Freeport is not one of toxicity, it is one of sheer volume. By the time the Grasberg deposit is mined out, a billion and a half tons of sediment will have been transported by the Aikwa to a tailings impoundment area.

"The issue is the *amount* of tailings," says Dr. Wisnu Susetyo, who heads Freeport's $2.5 million environmental monitoring lab in Timika. Although his Ph.D. in chemistry is from the United States, Susetyo comes from Yogyakarta in Central Java. He explains our situation by way of an analogy: "If you keep two or three chickens, it's fun, it's good for you, and you have eggs to eat. Two or three chickens are not difficult to feed, and you can find a place to keep them. But say you have *five hundred thousand* chickens—then you will have a problem. How do you house them? Where do you house them? How do you feed them? How do you control the noise, the smell, all sorts of things? This is the issue with the tailings."

The potential problems created by a river with such a high sediment load were dramatically revealed to us after a particularly violent storm in the middle of 1990. The high winds and floodwaters brought about by this storm washed trees and brush into the Aikwa, causing its channel to block up. As water continued to build, the silty river overtopped its bed, and began to "sheet flow" eastward. The uncontained river flowed in a broad, but shallow sheet through a forested floodplain, finally reaching the Kopi River, part of the neighboring Minajerwi watershed. (See MAP 10.1, page 269.)

Once the floodwaters subsided, the event had turned the already wide and braided channel of the Aikwa, beginning where the Otomona and Aikwa meet just above the Timika area, into a broad band of sediments extending south past the airport and hooking east to meet the Kopi River, and the Minajerwi River just south of where it joins the Kopi. The muddy plain that was left is in places five kilometers wide. The sediment buried shorter vegetation, and together with the standing water, cut off the oxygen supply to the roots of the trees growing there, stressing and killing them. Freeport engineers designed and built a levee system on the western edge of the Aikwa's old channel to protect the population of Timika and the road to the port, but the river never returned to its old path.

The mountains of Irian Jaya are rugged and criss-crossed with faults, and erosion is proceeding there at a very rapid rate. The sediment load in all the rivers along the island's south coast is very high, and flash flooding and river course changes are regular events. Comparing good World War II–era Dutch maps to current satellite images, one can see that, while upland river courses have remained the same, in just the last fifty years the lower runs of rivers and river mouths have been noticeably reshaped by sediment deposition. Flying over the south coast, far from our mining area, one often sees dead trees caused by choking sedimentation, and in some areas the sediment, freshly unearthed by flooding, has the same gray color as the Aikwa.

Still, there is no doubt the sheet flow of 1990 was at least exacerbated by the high sediment load in the Aikwa from Freeport's mining activities, and the very visible results of this flooding helped focus the company on solutions, and finding a management plan for the tailings. In the mid- to late-1980s, the company was facing just 10–15 years before the ore bodies would be exhausted and the mine shut down. The Indonesian government authorities unofficially urged Freeport not to establish "permanent facilities" in the area for fear they would be forced to bear the continuing expense of maintaining them when mining stopped. A temporary philosophy prevailed. Then the Grasberg deposit proved out, and everything changed.

We hired Bruce Marsh as environmental manager in 1991, and his charge was to start a full-time, on-site environmental department. Marsh, boyish and articulate, comes across as a bit collegiate—the polar opposite of a standard-issue hard rock miner. This did not help him in trying to work with our on-site operations personnel, who seemed to take great pleasure in providing this "silver spoon boy" with some proper instruction in the real world.

The support he received from New Orleans headquarters helped, but basically Marsh was interested in doing the job he was hired for, and if he had to step on some toes, well, he just made sure he did it nicely. One of his first tasks was to try to figure out how to manage the tailings. The sheet flow was carrying most of the sediment load from the Aikwa system into the Minajerwi system. Since there is naturally a difference in grade between the Aikwa and the Minajerwi, over time the Aikwa might well cut a more direct channel to the neighboring river. Although precise figures are very difficult to determine in such a little-studied and changeable environment, Freeport consultants estimated that in this case, a large percentage of the Aikwa's load could shift to the Minajerwi. If this were to become the case, computer modeling predicted that the Minajerwi could plug and overflow its banks in as little as 11 years.

Irian Jaya's mountains are young and steep, the area is seismically active, and a large volume of rain falls. The naturally eroded material produced by these conditions and entering the rivers is what a geologist would call "poorly sorted"—boulders, cobbles, gravel, sand and finer sediment. The larger aggregate is deposited quite quickly, and only the sand and finer material makes it as far downstream as Timika. The finest silt and clay stays in suspension all the way out to

The tailings-laden Aikwa River. Though it carries a greatly increased sediment load, its chemistry and biology are not fundamentally different from other area rivers.

the Arafura sea, and from an airplane one can see brown fans extending kilometers out from the river mouths.

The tailings produced by the mill are, in comparison, of uniform size—"well sorted," in geologists' parlance. Thus, they flow further downstream, and are deposited in a much more restricted area. The mill, at 2,800 meters elevation, discharges the tailings into the Aghawagong River, which flows into the Otomona Timur ("East Otomona"), which flows into the Otomona River, which flows into the Aikwa River. This is the point at which most of the tailings settle out, although the flow eventually reaches the Mukumuga estuary and the sea. Since the 1990 flooding, the Aikwa's flow has crossed over to the Kopi, and from there to the Minajerwi River. South of the sheet flow area, the old Aikwa channel is now virtually dry. The dry bed is now rapidly being taken over by volunteer vegetation and will probably soon be unrecognizable as having once been a river.

We hired consultants to analyze the various options for controlling the deposition of these tailings, and assembled a team of experts—Indonesians, Americans and Australians—to review the

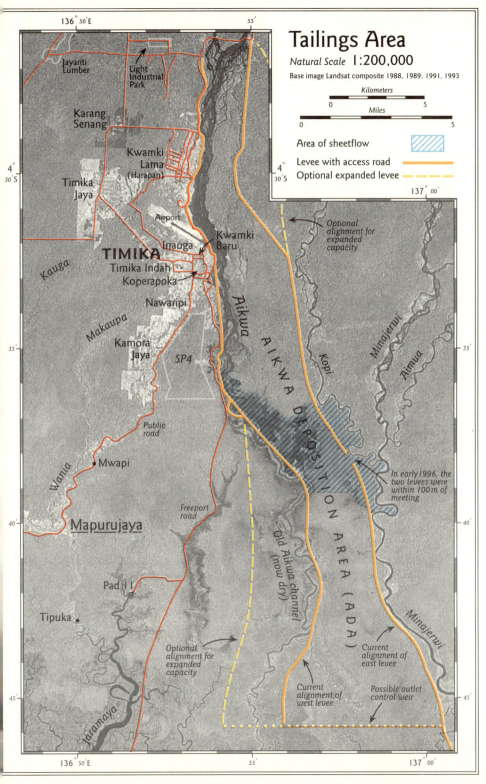

MAP 10.1

alternatives, including a pipeline to the sea, highlands tailings dams, lowlands dams, and diversions to the other rivers.

Some environmental groups wanted Freeport to build a pipeline, carrying the tailings directly out to sea. Fluor Daniel conducted a feasibility study on the pipeline, and their estimate was that it would cost $750 million to build, and that it would be extremely expensive and problematic to run. They could not even guarantee the technical viability of such a pipeline, since none like it exists anywhere in the world. Because of some intermediate elevation changes, the material would have to be pumped, which would consume a tremendous amount of energy. Also, the area is very earthquake-prone, and a sudden break in a pipe of such capacity could be disastrous. Finally, the Arafura Sea is uniformly shallow—even 20 kilometers out it barely reaches 50 meters—and there is no hole, or deep channel, into which we could deposit the sediment.

One of the primary concerns in whatever tailings management plan was selected was to avoid harming in any way the Lorentz Nature Reserve, a 1.5-million-hectare land area that begins two river systems east of the Freeport project area. The Lorentz reserve covers a broad swath of Irian Jaya from the mountains to the sea, and is one of the most biologically diverse and important reserves in the world. It is named for H. A. Lorentz, the Dutch explorer who in 1909 was the first European to explore this region, eventually reaching the snowfield of Mt. Wilhelmina. (Wilhelmina is now called Trikora, and no longer has an ice cap.)

We eventually opted for a system of levees with which we would, in essence, create a controlled flood plain, called the Aikwa Deposition Area or ADA. The western levee will be a continuation of the one already built after the 1990 flood to protect Timika. The eastern levee, three kilometers or more away, will contain the river, and when completed will keep it from entering the Minajerwi watershed, whence it might eventually be able to affect the Lorentz Reserve. The river will be allowed to find its own meandering, braided path within these two barriers, and the tailings will build up controllably in this plain. The ADA, which gained the acceptance of the Indonesian government, has enough capacity to accommodate future expansions, and has enough flexibility to keep up with Irian Jaya's fickle rivers—building the levees will be an on-going process, and their height and sweep can be altered as needed.

Eventually, the levee system will enclose 133 square kilometers. Some 49 square kilometers of this is area already affected by the sheet flow; the other 84 square kilometers will come from an area of swamp south of the sheet flow, which eventually can be covered in tailings. This swamp contains some sago palms and mangrove trees, but we considered this a better compromise than directing the tailings to the upland rain forest further north, or not containing the tailings at all. The levee system will cost $25 million to initiate, and about $12 million a year to maintain.

Even before the end of the mine's life, the deposited tailings will be a resource. They will have created new, usable farm land where a swamp existed. Also, if the future offers efficient enough techniques to remove the minerals still contained in the tailings, they can continue to be a mining resource.

The mill tailings from the Freeport mine, spreading in the artificial floodplain created by the levees, are responsible for the most visible environmental impact created by the mine. But the overburden, the barren rock that needs to be removed to get at the rich ore, presents another challenge to the environmental department. A billion and a half tons of tailings will be discharged by the mill over the mine's lifetime, which seems like a lot of sediment, but even before this rock gets to the mill, almost three billion tons of overburden will have to be moved aside.

Near the beginning of the mining operations at Grasberg, we hired scientists to survey the ecology of the area. New Guinea's ecological zones tend to be based on elevation, and the biological makeup of the Grasberg area was the same as areas east and west of the site, at the same alpine elevation. The scientists found no plants or animals there that were unique to the mining area. They collected seeds of all the species of plants found on Grasberg and Carstenszweide, and planted them at a botanical laboratory at the mine site. This is our ark. Once mining is completed, the whole area will be resown in the same species that existed before the stripping began. Conditions at the elevation of Grasberg are harsh, and the natural vegetation found there is hardy and not luxuriant; but our tests have shown that it will propagate into the freshly disturbed rock.

The problem with the overburden is not replanting. The problem comes from a phenomenon known as acid mine drainage. An overburden pile, made up of boulders, cobbles and smaller rocks, is

The Carstensz glaciers in 1936. Note the uninterrupted wall of snow presented by the Northwall Firn in the background, and the Meren and Carstensz Glaciers.

porous, and exposed to the air. If sufficient quantities of sulfide minerals are present, the oxygen in the air, combined with the area's heavy rainfall and naturally occurring bacteria, can bring about acidification. This acid can enter the watershed, with deleterious effects on the plant life, and it can also leach copper from the rock.* Whether or not an overburden pile has the potential to go acid depends largely on the chemical makeup of the rock. There is so much limestone in the Grasberg area that in the beginning of mining, the company assumed there could be no problem with acid mine drainage.

And fortunately, most of the Grasberg overburden does not generate acid. About 25 percent of the waste rock is well-buffered, and does not generate any acid—some of this has enough of a limestone component to buffer other rock. Another 25 percent has only a very low acid-generating capacity. The other half of the overburden rock has some measurable capacity to produce acidic water, and maybe

---

*Some mines are able to use this process—called heap leaching—as a method of extraction, thus recovering a resource from their waste (see previous chapter). Unfortunately Grasberg's mineralogy is not conducive to heap leaching.

The Carstensz glaciers in 1995. The glaciers have lost 70 percent of their area over the last half century, and have now broken up into six separate ice fields.

one-fifth of the total has high acid-generating potential. It's this latter that is the problem.

The overburden is being placed in two areas, Carstenszweide and the Wanagong Valley. Eventually, 114 hectares of the meadow will be covered 250 meters deep, and 769 hectares of the Wanagong Valley will be covered 450 meters deep. By judiciously blending waste rock of various acid-generating potentials and, at times mixing in limestone, the chances of acid drainage are greatly reduced. Further, the natural sediment in the area's streams is chiefly limestone, and their buffering capacity is thus quite high.

Even after mixing the different grades of overburden, and perhaps adding limestone, in places the waste piles may still have acid-generating potential. You can't protect the overburden from water, but you can protect it from oxygen. A layer of clay or topsoil, and up to 20 meters of inert rock, provides a very effective seal over the problem rock. A variety of strategies will likely be used, as the overburden piles grow over time, and as our ongoing monitoring and analysis programs recommend.

ANY OPEN PIT mining operation creates a very visible impact on the local environment, but because of our unique location, we have been blamed for some things that our operations have had nothing to do with. The shrinking ice fields a few miles to the east of the mine site are perhaps the most striking of these. The favorite story around the company was the report in Jakarta that our Heavy Equipment Access Trail was causing the glaciers to melt, a charge supported only by the road's colorful acronym: HEAT.

Not all the charges were this ludicrous, however. Trekkers who climbed the glacier noticed tiny dark spots in the ice, like soot, and charged our mine operations with having caused this "pollution." Further, they suggested that our blasting operations at Grasberg were destabilizing the ice. We decided to call in experts to study the Carstensz glaciers, to find out exactly what was happening, and if our operations were having any deleterious effect on the ice fields. We flew Dr. Alex W. Wilson, who has researched glaciers at various places in the world, to Irian Jaya. His discovered that the "soot" was in fact cryoalgae, a type of naturally occurring blue-green algae that can grow in ice, and is found in virtually all of the world's glaciers. Its growth is enhanced by the extremely clean conditions of the Carstensz glaciers, which, being in quite rapid retreat, have not accumulated the dust and debris on their leading edges that would otherwise block the light necessary for its growth. Other consultants placed sensitive monitors on the glacial ice to determine if any seismic or airblast effects from the Grasberg mine were reaching the glacier. Their conclusion was that these effects, compared to the background seismic activity in the area, were insignificant and mostly unmeasurable, and in fact that a single trekker walking on the ice created a much greater effect.

Although the Grasberg mine is not to blame, the glaciers of the Sudirman Range are definitely shrinking, as are tropical and temperate glaciers around the world. The reason is a natural global warming trend of the last 130 years, possibly accelerated recently by the so-called "greenhouse effect." The world's rapid industrialization, accompanied by the burning of fossil fuels, has sharply increased the percentage of carbon dioxide in the atmosphere, which causes heat to be trapped, as in a greenhouse, raising the earth's average temperature. Some scientists estimate that by the middle of the next century, today's carbon dioxide level will have doubled, bringing a temperature increase of between two and four-and-a-half degrees Centigrade.

During the last ice age, which lasted from 125,000 to 15,000 years ago, the Carstensz glaciers extended 17 kilometers down the mountain. At that time Tembagapura's neighboring village of Banti would have been covered in ice. As the natural warming cycle progressed, the ice retreated, until by about 7,000 years ago there was no ice at all. The glaciers re-formed during what scientists call "The Little Ice Age," a period of global cooling lasting from A.D. 1350 to A.D. 1850. It was at the height of this period, in 1623, that the Dutch Captain Jan Carstensz, sailing by the south coast of the island, first sighted the snows on the Carstensz peaks.

By 1865, the glaciers were once again in retreat. The first expedition to measure the extent of the glaciers was the Colijn expedition in 1936, during which Jacques Dozy noticed Ertsberg, Grasberg, and most of the other geological features in the area. The massive erosion of the last ice age is what revealed Ertsberg. The ice carried away the softer rock around it, but the hard, igneous composition of Ertsberg left it exposed to Dozy's trained eye. And Grasberg stood high enough to avoid the scraping of these rivers of ice, one reason so much of its rich mineralization remains.

In 1972, a group of scientists from Australia conducted a detailed study of the glacier, comparing contemporary satellite images and maps of the glacial extent to Dozy's 1936 maps, and to the position of the stone cairns he left to note the edge of the ice fields. The retreat between 1936 and 1972 was substantial. When Colijn and Dozy climbed these mountains, the ice pack was such that Ngga Pulu, snow-covered and on the edge of the Northwall Firn, was the highest point in Southeast Asia, a distinction that now belongs to Puncak Jaya. In 1936, Dozy's corrected barometer reading for Ngga Pulu indicated 4,906 meters; the 1972 expedition recorded just 4,862 meters. Puncak Jaya, too steep to hold snow, has always stood at 4,884 meters. In 1936, 13 square kilometers of the Carstensz mountains were covered in ice. By the early 1970s, only 6.9 square kilometers remained covered. (See MAP 10.2, page 277.)

Three other mountains in Irian Jaya have carried snow in recent times. In fact, the race to the snows of Dutch New Guinea was won by Lorentz, when he and his Dayak carriers reached the snowline of Puncak Trikora, then called Wilhelmina Top, some 170 kilometers east of the Carstensz ice fields. In 1909 Lorentz reported that the ice

The hardy swamp grass Phragmites *sends out rhizomes that quickly colonize land created by deposited tailings. The grass stabilizes and enriches the tailings.*

extended to 250 meters below the summit. Nobody knows exactly when this ice disappeared, but by the 1970s Puncak Trikora was bare. Gunung Mandala, the highest point in the Wisnumurti Range near the border with Papua New Guinea, also was reported to be snow-capped, but no longer. As recently as 1990, a small relict ice field could still be seen on Ngga Pilimsit, also called Idenburg Top, just west of the Freeport mine area, but by 1993 it also had melted.

Today, the ice-covered area around Puncak Jaya, including the Northwall Firn, the Meren glacier and the Carstensz glacier, has shrunk to less than 4 square kilometers. The equilibrium line, above which there is active snow formation, now lies just 100–200 meters below the highest peaks. In 1994, only the Carstensz glacier and the eastern edge of the Northwall Firn showed modest signs of active glacial ice flow. In just the last eight years the Meren Glacier, which once flowed like a river down the Meren Valley, has separated from the Northwall and become a rapidly shrinking firn, or ice field. Within ten years it may be gone, and within twenty-five years it is possible all the Carstensz snow fields will have disappeared.

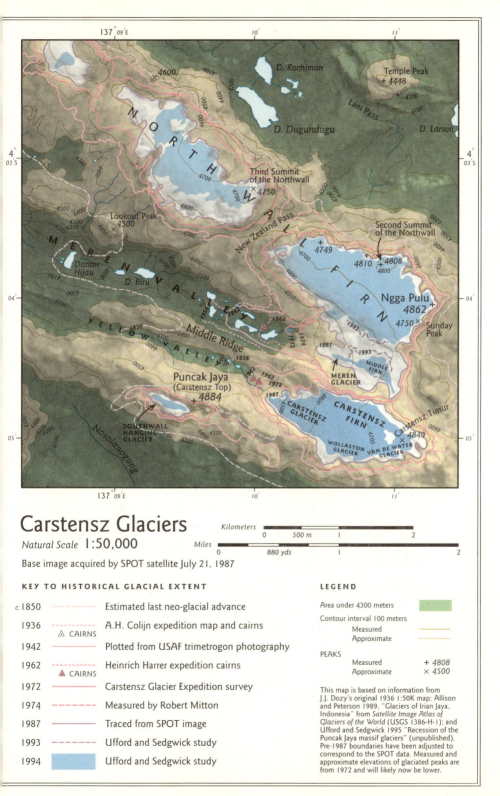

MAP 10.2

From the barren, 4,000-meter highlands of the glaciers to the dense lowlands, the area around the Freeport mine and operations includes some of the most interesting terrain, and wildlife, on earth. New Guinea is the world's largest tropical island, and its diverse fauna and flora includes the largest and smallest parrots on earth, the biggest doves, the smallest frogs, the largest butterflies, the longest stick insects, the highest tropical trees, the largest rhododendrons and the largest remaining mangrove swamps. The biota is a complex mix of species that came from different directions: north from Australia, and east from Asia.

A land connection between New Guinea and Australia across what is now the shallow Arafura Sea existed until the end of the last ice age, just 15,000 years ago, and like Australia, New Guinea's mammalian land fauna consists principally of marsupials. Bats, rodents and some domestic placental mammals have either flown to the island or accompanied human beings. The island's wildlife are still rather poorly known, and as recently as 1994, Dr. Timothy Flannery, an expert on Australian and New Guinea mammals, and Dr. Boeadi, a Bogor-based veterinarian, identified a new species of mammal here.

To discover a new mammal anywhere in the world at this late date is a rare event indeed. When reports of a new tree kangaroo began coming in, Freeport lent logistical support to the scientists to help them collect and make a positive identification of the new animal. This unusual animal, whose markings make it look like a panda, was dubbed *Dendrolagus mbaiso*. The species name comes from the Moni word *mbaiso*, which refers to an animal that is sacred and cannot be hunted. This is fortunate for the animal, also called *bondegezou* or "man of the forest," because east of the Moni in the Dani areas there are no such restrictions and the almost tame tree kangaroo has been hunted nearly to extinction.

The island of New Guinea, as a whole, has fewer plant species than other tropical areas of similar size, even those with less diversity of habitat. But a great number of these are only found in New Guinea, perhaps 55 percent of the 9,000 species so far identified. Because of the cold and clouds, the botany of the alpine region around the mountains and mine operations area is generally poor, consisting of a few hardy sedges and shrubs. Some stunted trees cling to the steep slopes, their bark covered with lichens, liverworts and mosses. A little lower, the vegetation gets thicker, but the highest region supports only a few, hardy species. Below about 3,800 meters, the mixed, broadleaf forests begin. At the higher elevations, where the clouds form a perennial blanket, the trunks and branches are draped with moss. As you move lower, conifers appear, and the canopy is higher. Palms and liana vines become more frequent in the lower montane forest, where the canopy height averages 30 meters. By the time you reach the mixed lowland forest, the canopy is thick, and 45 meters above the ground. This plant association is the most complex and least understood on earth. Orchids and other epiphytes cling to the canopy, and feather palms and the distinctive stilt-rooted pandanus palm are common.

The high rainfall in this region produces a network of meandering rivers, each major system just a few kilometers apart. As they near the coast, the rivers become a bewildering labyrinth of interconnected channels. The Kamoro people of the coast can guide their canoes for tens of kilometers along the coast without ever having to go out to sea. River life is only present in the lowlands here, and at elevations above a few hundred meters the streams are barren. The nearly daily deluge of afternoon rainfall flushes them thoroughly, and New Guinea's brief history as an island means that the fish species here are all relatively recent evolutionary departures from saltwater species. Varieties suited to fast, cold highland streams are still absent, not having had sufficient time to adapt.

The huge swamplands along the coast harbor valuable tropical hardwoods and pandanus, as well as sago palms. The starchy interior of the sago palm's trunk serves as the staple crop of New Guinea's coastal people. Mangroves invade and dominate most brackish stretches at the edge of the murky Arafura sea, although on sandbanks and drier ground, feathery casuarinas and hibiscus grow.

Wildlife studies by Freeport have focused on habitat, as it has been well established that diverse plant cover is the key to maintaining a rich community of wildlife. In areas where the company's operations may potentially disrupt this community, "greenbelts" or "wildlife corridors" have been established. For example, greenbelts are a key part of the design and layout of the new town of Kuala Kencana, built by Freeport north of Timika.

With all the road building and other construction taking place, many areas become stripped of vegetation, leaving ugly gashes in the landscape. Not only is this aesthetically displeasing, but the removal

Areas cleared in the course of construction are revegetated by spraying a mixture of seeds and fertilizer on the bare soil, an efficient process called 'hydromulching.'

of plants, with their soil-binding roots, leaves these areas prone to landslides. Freeport revegetates these disturbed areas using a technique called hydromulching. Seeds from our stockpiles of native species are mixed with fertilizer, peat and a binding agent. This mixture is loaded in a tanker truck, and sprayed at high pressure on the bare soil. Very soon a velvety green layer appears, and the roots begin to fix the soil in place. The procedure has been very successful, and areas hydromulched a few years ago are indistinguishable from the surrounding vegetation. When our work in Irian Jaya is finished, the unneeded roads and construction sites, the overburden area, and the upper benching of the Grasberg pit will all be hydromulched.

T HE TAILINGS DEPOSITED by the Freeport mine end up in the zone between the area of lowland forest and the beginning of the sago swamps. Already Phragmites, a hardy and aggressive swamp grass, has colonized large areas of the deposited tailings. The pioneering swamp grasses are followed by bushes, and as the tailings are stabilized and lightened with the natural organic matter, finally trees.

Savannahs of Cyathea tree ferns are a distinctive feature of Irian Jaya's upper montane zone, growing in well-drained areas above about 3,000 meters.

When you pick up a handful of tailings, it would seem that nothing could grow from the fine, sand-like material, but the natural recovery began so quickly that Freeport scientists started experimenting with farming the tailings. We are in the process of testing corn, pineapples, grasses for livestock, nut trees, coffee, cacao, and even rice, along with other potentially useful plants. The first demonstration project, begun in 1992, covered 66 hectares, and today we are working on more than 250 hectares, which is all the recoverable land so far created. The results have been surprisingly good. Our agricultural scientists have shown that a wide range of plants will not just grow, but actually thrive on plain tailings. The substrate gets better with each season, a process that can be accelerated with the addition of a small amount of organic material or waste sewage sludge.

The area recovered by the tailings management plan will eventually include 133 square kilometers, which will be built up gradually between the two levees. In addition to allowing natural vegetation to grow, long-term reclamation plans for this area include agriculture, horticulture, silviculture and aquaculture projects. It will take some experimentation over time to determine the species and sequence of planting that will be most suitable and which will yield the best economic results. Our farming station employs sixty local Kamoro men. They now work the soil for the experimental plots, and eventually will show their fellow coastal farmers how to effectively grow staple and cash crops from the reclaimed tailings.

In July 1992, the first 66-hectare plot was set aside at Mile 23, on the west side of the Aikwa River, just south of Timika. We avoided any solutions that require expensive materials or complicated technologies, instead relying on simple, proven methods familiar to the local farmers. When we analyzed the tailings, we found them to have more phosphorous and other plant nutrients than the limestone soil of the area, but to contain virtually no organic matter or nitrogen. Nature was already improving the situation by colonizing the barren tailings with *Phragmites*. This vigorous plant grows up to four meters high, and although it does not fix nitrogen like a legume, its substantial biomass adds considerable nitrogen to the soil. Questions researchers are still exploring: How long should *Phragmites* be allowed to grow before switching to a farmed crop? Can another species add nitrogen more quickly or efficiently than the swamp grass? Can small additions of soil bypass the need for a green manure stage?

The initial plot at Mile 23 proved less than satisfactory for experimentation, because the water table there was just 20 centimeters below the surface, and it was prone to flooding. So another plot was started at Mile 21, a bit further south. Here the layer of tailings is fairly deep, and the water table lies two meters below the surface. The thick, well-drained layer of tailings here more closely approximates the conditions that will exist in the ground recovered by the tailings management plan.

One of the most unlikely crops to plant in the tailings is irrigated rice. Rice paddies require enough stability in the soil to maintain the small dikes that separate the ponds, and the raw tailings slump easily like fine sand. Although not a local Irianese staple, the Indonesian population of Timika has an ever-growing need for rice, which is currently being filled by imports from Java and Sulawesi. There are transmigration farms in the area, but their land is not suited to rice. Growing rice locally, especially in tailings, would fulfill a very pressing need, and Freeport agronomists have set up a number of experimental plots to see if this can be accomplished.

Howard Lewis, the Freeport environmental department manager who oversees the reclamation program, said the first problem was keeping the dikes, or berms, from immediately slumping and becoming useless.

> We looked at armoring the berms with *Phragmites* grass. We cut it, put it into bundles, and then laid the bundles over the berms separating the ponds. But the *Phragmites*, even after it was cut and dried for a while, still sprouted from its nodules. So it grew right up. This was my idea. I thought, we've got all this material, it grows several meters tall, let's cut it and bundle it and apply it to the banks. But boy, that didn't work. Even after drying it out for several days—or even a week—it still came back and formed a rooted plant, and we got *Phragmites* in all our ponds.

The Kamoro laborers working for the department came up with a better idea—sago trunks. This tree, which yields their staple crop, grows in nearby swamps, and they knew its trunk was very resistant to deterioration. So they cut down some trunks and used these to line the paddies. This experiment was successful and allowed us to create some trial paddy fields. Some of these have yielded good results, but it is still too early to tell if large-scale production is possible. Recently, this project was discontinued in favor of less labor-intensive dry land techniques for growing rice on the tailings.

As part of our reclamation program, we have also created a river and a series of small lakes near Kwamki Lama, a settlement north of Timika town, by redirecting some of the small streams in the area. The lakes principally serve for beautification and recreation, although another series of lakes created by Freeport, further downriver, are full of fish. Driving by these lakes, which cover an area of 500 hectares near the airport, one can often see people casting their nets from small rafts. Women wash clothes along the shore and children splash in the shallows. Local communities occasionally organize canoe races on these lakes to celebrate festivals, and holidays such as Indonesia's Independence Day.

Tourism to the Timika area shows potential. We consider this an opportunity for the future, and we built, and then privatized, a four-star hotel adjacent to the airport. The Timika airport itself was built by Freeport more than 25 years ago. The new four-star Sheraton Inn Timika, which opened in late 1994, provides a comfortable stopping place for visitors. Forty hectares of land behind the hotel has been set aside as a bird sanctuary. The land had been somewhat disturbed by local inhabitants, but today it is rapidly revegetating. Recently, an expert visitor identified 91 varieties of birds in five days, including 6 different species of birds of paradise.

WITH MORE THAN 17,000 people and a billion dollars worth of machinery working in the area, a lot of waste is generated. Some of this is harmless, causing more of an eyesore than any real environmental damage. Other waste is more potentially harmful. But all these wastes have to be carefully managed, and they must be disposed of in an ecologically sensitive manner.

When the shift breaks, and a couple dozen bone-tired underground miners emerge from the portal, enjoying their first smoke of the day in the open air, who is going to tell them they can't throw their butts on the ground? Many of our employees do not have the heightened awareness of litter that Americans do—and of course, it took two decades of education, reinforced by heavy fines, to break the American habit of throwing trash on the ground. (In Louisiana, where I now live, it seems as if it may take another 50 years.) Indonesia is just beginning this education process, and it will take time to carry through. Our managers must continue to become more aware of their responsibility to the environment. Even with the most efficient environmental department and good waste disposal programs, success requires continuing education. Bonus-generating production targets in our operations now take the environment—as well as safety and hygiene—into consideration.

Even the best efforts sometimes fail. Freeport has for years been trying to get employees to properly dispose of the aluminum foil wrappers from their take-out meals. All the workers in the mine area, and many of the survey and exploration workers, are too far from the company mess halls to take their mid-shift meals there. They take their food from one of the self-service mess halls, packed in aluminum foil. This foil is expensive, and is an eyesore when it ends up as litter. But when the mine management decided to substitute a biodegradable wrapper for the take-away meals, it was faced with some very angry workers. Only food wrapped in aluminum foil could be easily reheated over a simple burner, and a cold meal is not what a miner wants after half of a hard shift. Management was forced to return to foil. The environmental department has had a lot more success persuading housewives to switch to phosphate-free detergents.

From a strictly economic point of view, it might not make sense

KAL MULLER, 1995

Stacked into piles and filled with concrete, recycled tires strengthen a section of the roadway. Freeport vehicles wear out 1,000 tires each month.

to recycle much of the waste generated by the mine operations. The cheapest solution, taking the high costs of manpower and transportation into account, would be to simply dump it somewhere nearby. We can import most of what we need duty-free, but can sell these items within Indonesia after use only if full duty is paid. This restriction applies to materials which could perhaps be recycled in Jakarta or elsewhere in Indonesia. This constraint makes recycling difficult. In this atmosphere, the environmental department has had to come up with some clever solutions.

Today, Freeport recycles all light vehicle and transport truck tires, which wear out at the rate of 1,000 a month, all the recoverable waste oil, tram cables, and a good percentage of aluminum, copper wire, scrap steel and some construction materials. We have even started composting some of the mess hall wastes. At the moment, we are working toward an agreement with the government to able to ship recyclable materials for processing in Indonesia.

For several years now, used tires have been collected and used to reinforce roadsides. Filled with gravel or concrete, they are effective at preventing erosion from the normal heavy rains. Used cables from the aerial tramway serve as guard rails in the more dangerous areas of the road. Used tires bought in Indonesia can be shipped back to the Jakarta supplier for credit, and the same applies to used oil drums and wooden pallets.

The biggest economic success coming from recycling is the oil recovery program. With all the heavy machinery and vehicles at the mine, some 175,000 gallons of lubricating and other oils are used each month. About 15 percent of this can not be recovered, because of leaks and transfer losses, and residues left in filters and lines. But that still leaves a lot of oil, which we gather up with a storage tanker and vacuum truck pickup system. In its first year of operation, the tanker recovered a monthly average of 44,000 gallons, and by last year we had reached 100,000 gallons a month, thanks to better storage systems and environmental awareness. We have a use for all this oil—it is burned as fuel in the port site concentrate dryers. Our requirement last year was 106,000 gallons of fuel oil a month, which cost us $75,000 each month. By burning waste oil in our kiln, we can save $900,000 a year. So, in this case, what's good for the environment is also good for P.T. Freeport Indonesia's economy.

Some of Freeport's scrap metal ends up at the blacksmith shop, where it is hammered into *parangs*, the Indonesian machete, shovel heads, hoe heads, and other agricultural implements. This is part of the business incubator program, in which local Irianese are given the facilities and training to produce useful goods or services. An Irianese man runs the bellows, while his partner alternately heats, and shapes an old truck spring. Within a few minutes, a strong spade head begins to form. With so many transmigrant farmers in the area, local demand for such implements is very high, and the spade will be sold just about as soon as it is quenched. These locally made tools sell for about half the price of imported tools in the Timika market, and their strength and quality are usually much higher.

Although a very successful program, blacksmithing cannot take care of all the scrap metal. We are looking at the possibility of shipping some to Australia for recycling. Another still unsolved problem is used batteries. The mine operations wear these out at the rate of 400 a month. Recovering lead from these batteries in an environmentally safe way requires the kind of operations that exist in Australia, or the United States, but unfortunately not in Irian Jaya. The environmental department is still working on a solution to battery recycling. Aluminum cans are collected and crushed, and together with whatever foil can be recovered, are shipped to Jakarta for recycling. A containerload of this material, its three to four tons representing about one-fourth of the total aluminum used by the operations, leaves Irian every three months. As the company cannot take any of the money from the recycling, it is put into a special fund for civic activities, such as for girl and boy scouts.

Finally, whatever cannot be recycled ends up in a carefully selected and managed final storage site. This site, prepared in the lowlands, is in the process of passing into private hands as part of the overall privatization program. But Freeport specifications must be followed as to proper design, construction and maintenance of any waste disposal sites. This includes a specially designed incinerator for medical wastes, which takes care of waste material from the Tembagapura hospital.

The company's concern with the environment gave birth in early 1994 to a modern, state-of-the-art, $2.5 million scientific laboratory in Timika, headed by Dr. Susetyo. The quality and instrumentation of this lab is matched in Indonesia only perhaps by one Jakarta laboratory, and many high-end labs in the United States would envy Susetyo's two brand-new Perkand Elmer graphite furnace, atomic-absorption spectrometers. These machines analyze samples to the parts-per-billion level.

The environmental lab has established 25 strategically located monitoring stations along the Aghawagong–Aikwa tailings drainage and neighboring river systems, plus 16 additional stations around the mine. Water samples, sediment samples, and samples of fish and shrimp are regularly collected from these stations and subjected to the scrutiny of Dr. Susetyo and his highly trained Indonesian staff. The retinue of tests examines 30 different parameters, including heavy metal content and major ions. So far, all water and biota samples conform to or surpass the most stringent environmental standards of Indonesia, as well as the published drinking water quality criteria of the World Health Organization and the United States Environmental Protection Agency. For example, Dr. Susetyo says water at the Otomona bridge station regularly tests at less than .01 parts per million of copper. The Indonesian government standard is 1 part per million, some 100 times higher than the Otomona, and the World

Freeport biologists seine for the river shrimp Macrobranchium. Shrimp from the tailings-laden lower Aikwa are just as healthy as those from neighboring rivers.

Health Organization standard allows 2 parts per million, some 200 times higher.

Freeport started its long-term environmental monitoring program (LTEMP) in 1990, and since then has built up a substantial database. The LTEMP is formally divided into five categories: meteorology, hydrology, water chemistry, sedimentology and biological sampling, as well as special field and laboratory projects. A compilation of all the results is, at three month intervals, presented to relevant authorities at both the central and provincial governments.

The climate within Freeport's area is monitored by six permanent meteorological stations, which continuously record air temperature, relative humidity, wind velocity and direction, solar radiation, and rainfall. Three hydrological stations record river dynamics by measuring the height of the surface water with ultrasonic sensors. Both types of stations operate 24-hours-a-day, with their power provided by solar cells and battery storage. Periodically, the data they collect is downloaded into computers and analyzed.

Biological analysis centers around heavy metals present in animals found in the river system. While a variety of species is tested, two have been taken to be the key long-term indicators: the catfish, *Cochlefelis spatulate*, and the river shrimp, *Macrobranchium* spp. These two were chosen as they are both bottom dwellers, thus consuming food from sediments, and both are regularly eaten by members of the local community. Because it is a crustacean, and thus has copper-based blood chemistry, the shrimp takes up more copper than the catfish, but in neither case has the animal been found to have dangerous, or even unusual levels of copper.

In most cases, the amount of copper that ends up in the fish, shrimp or plants depends on the particular species' tendency to absorb it into its tissues. The rice growing on the tailings has an average copper content of 3.3 parts per million, well within the Australian National Food Authority's 10 parts per million standard. (Indonesia does not yet have standards for heavy metal content in food plants or animals.) In comparison, rice growing on the transmigration farms, far from any impact of the tailings, tests at 3.1 parts per million of copper. This is slightly less, Dr. Susetyo says, but the transmigrant farmers also grow a different variety of rice. Dr. Susetyo and the laboratory staff happily eat local fish and produce. In fact, he says, he would rather eat fish in Timika than in a Jakarta restaurant, as the latter can have high mercury levels.

Freeport also runs a very important biological sampling and species diversity program. Our scientists regularly collect from both the Aikwa drainage and the other nearby rivers unaffected by our tailings discharge, the latter to serve as a control. They have found no significant differences in the types and numbers of species living in the Aikwa and in the other south coast rivers. Our gray river, despite its obvious difference in color, seems no less healthy, or rich in animal life, than Irian Jaya's other south coast rivers.

CHAPTER ELEVEN

Community Relations

A special obligation—The Kamoro—Missionaries and the Dutch—The Amungme—Cargo cults—Land ownership in Irian—Health care—The business incubators—Expectations—Sustainable development

IN 1969, WHEN Freeport first began constructing the roads, port, town, mill and other infrastructure that would be needed to mine Ertsberg, the local population in the area was sparse and scattered. A few hundred Irianese of the Amungme ethnic group lived in the forested highlands, planting sweet potato vines on the steep mountainsides, raising pigs, and hunting for game. Along the coast, a few hundred Kamoro lived in shifting hamlets along the rivers, netting fish and prawns in the tidal rivers, hunting for cassowary and wild pigs in the lowland forest, and harvesting the starchy sago palm.

The only populated areas near the Freeport operations were the Wa Valley, a few kilometers west of where Tembagapura now stands, where perhaps 50 or 100 Amungme lived, and a semi-permanent fishing hamlet of fewer than a dozen Kamoro, near what is now port site.

Today, 50,000–60,000 people live in the area around the Freeport operations, and it is one of the most developed and populous areas in the province. The bustling town of Timika, with 35,000 people, a jet airport and a new four-star hotel, stands on what was an empty stretch of flat land west of the Aikwa River. Tembagapura, a modern company town housing 17,000 people, dominates a steep-sided highland valley that just thirty years ago was empty.

There is no doubt that the Freeport mine has had a profound impact on the area. The population of the original Amungme, Kamoro and Sempan people in the area has increased moderately to about 2,500 highland Irianese and about 4,000 Kamoro and Sempan, and the mining activities have attracted thousands of Irianese from elsewhere in the province—Ekagi from the Paniai Lakes, Moni and Dani from the central highlands, Biakans from Biak Island.

But it has been Indonesians from the more populous and developed islands of West and central Indonesia that have made up the bulk of this population increase. Javanese, Balinese, Ambonese, Bugis and Torajans are well-positioned to take advantage of the opportunities presented by our Irian Jaya operations. They come to Irian with a good basic education and prior work experience, and they have the skills to run the businesses that have sprung up in Timika, and to be hired at the mine.

The indigenous Irianese are at a marked disadvantage in competing for these jobs. Compared to the Islamic and Christian cultures of western Indonesia, the Melanesian cultural traditions of Irian Jaya are a poor match with the demands of modern capitalist economies. The jump from shifting cultivation of yams and gathering of sago to wage employment servicing truck transmissions is a long one. Outside of a few mission schools, there was no education here before the arrival of Freeport and the Indonesian government in the late 1960s. There was no basic health care. For these reasons, in its development work Freeport feels a special obligation to the indigenous Irianese, especially the local Amungme and Kamoro.

Our coming to the area has altered and will continue to affect the lives of the Amungme and Kamoro people in this area. Thus we feel an obligation to work with them, and this includes working with the Indonesian government in their programs, in the local development of this area. We have this obligation because we're in this for the long term. It also makes economic sense to hire locally and avoid the high transportation costs associated with workers from elsewhere in the three-thousand-mile archipelago.

Our various community programs emphasize a tripartite relationship between the local people, the government and Freeport, a relationship that is crucial if the programs are to be sustained over the long term. Freeport's many community activities include programs to address basic needs, such as health, education, and housing, as well as those to foster economic development and cultural preservation.

It is easy for an outsider to imagine the Amungme and Kamoro lifestyle before the mine operation began as an idyllic, Walden-like existence. However infant mortality stood at 50 percent, respiratory and gastrointestinal diseases were common, and some of the healthiest young men died in tribal warfare. In Freeport's first agreement with the people of the Wa Valley, signed in 1974, a clinic and a school

A group of Amungme men from Nargi hamlet, photographed by Jacques Dozy in 1936. Nargi was west of Banti Village, between the Wanagong and Opitawak.

headed up the Amungme list of requests. "A clinic to keep babies from dying" was how one Amungme leader put it. The advantages of education became obvious to the Amungme as soon as educated outsiders arrived.

Charlie White, our vice president in charge of sustainable development, said that when he first came to work in Irian, one of the most respected Amungme traditional leaders told him: "I would be a different man today if I had an education—I want my children, and my grandchildren, and all the children to have an education."

The contrast between their traditional ways and those of the modern world, into which the Irianese have been catapulted in less than one generation, presents serious challenges for planners of development programs. Clinic doctors, for example, still occasionally have to pull arrows out of men who show up at the door in feathers and war paint. Bruce Marsh, who heads Freeport's environmental department, tells one of the more interesting stories. Marsh from the beginning has hired local Irianese to tend the research gardens and work with various recycling and environmental testing programs.

An Amungme man living in Banti Village today. Freeport has aided the Amungme with health care facilities, education, jobs and housing.

These men, who were soon called "Bruce's Boys," are comfortable working in the climate, and possess indispensible geographical and biological knowledge.

> About a month into the job I was at the recycling center for my morning meeting. I had 'Bruce's Boys' all around me and we were talking about what we were going to do for the day. I took attendance every day, and that morning I noticed that a man named Namo wasn't there. So I said, 'Well, Namo's not here, so he's not going to get paid today.' And the supervisor of my men, Julius, turned to me and said, 'Mr. Bruce, Namo's here.' And I looked around again, and said, 'He's not here. Julius, if I don't see him, he's not here.' Julius lowered his voice and said, 'Mr. Bruce he's here. It's okay. You can check him off.' I looked around and said, 'Julius, do you see him right now?' 'No.' 'Then he's not here.' 'Mr. Bruce, he's here.' I said, 'Julius, he's not here.'
>
> Then Julius pointed and said, 'See those two guys over there?' And I looked over and here are two Irianese in kotekas [penis gourds], war paint, bows and arrows—I mean, they're dressed to the max. And I said, 'Yeah, I see those guys, Julius. Neither of them is Namo.' And he said, 'Well, they're looking for Namo.' And I said, 'They're looking for Namo? What happened?' And he said, 'Namo's father killed somebody—their brother or somebody like that—and they're going to kill

293

Namo to get back at Namo's father.' And I said, 'Julius, where is Namo?' And he said, 'You see that drum right in front of you? Inside that drum.'

So we're standing there around this oil drum, and Namo's hiding inside the drum, and I've got two guys over here in war paint, and I'm thinking, nobody ever explained this to me before I took this job. So we took the drum and loaded it into the pickup truck. I said, 'Take him to Timika until this situation cools off.' Then Julius and I went over and sat down with these guys. It took about six hours of talking, it took a couple of pigs, it took hundreds of thousands of rupiah, it took bags of rice, it took all kinds of stuff before they were happy and forgave Namo. We sent them back over the hill to Beoga, or wherever they came from, and we waited two weeks until we brought Namo back up from the lowlands to the environmental department.

IT WAS ONLY very recently that the people of Irian Jaya's south coast had any contact at all with outsiders. The Kamoro, living along the coast, enjoyed limited trade contacts, but the Amungme were almost totally isolated from the outside world.

The Kamoro- and Sempan-speaking people of the coast together make up a larger group known as the Mimika—the word is derived from one meaning a river that goes inland to fresh water. For at least several hundred years, the Mimika areas of Irian Jaya have been the easternmost extent of the important east–west archipelagic trade route through Indonesia. The tiny Gorong Islands, off the southeastern tip of Seram, and Dobo, in the Aru archipelago, were the main entrepôts of this eastern trade, which sought massoi bark, used as a medicinal in Java, crocodile skins, dried bird of paradise skins and, most lucratively, slaves. In exchange for these, the Irianese received metal tools, woven cloth, earthenware, and later, firearms. From the Chinese traders they received pottery and bronze gongs, which were used as bridal wealth.

Some 9,000 Mimika, including about 8,000 Kamoro and 1,000 Sempan, live along a 250-kilometer stretch of coast between Etna Bay and the Cemara River. The Mimika are coastal people and their settlements extend inland only a short way. Mimikan woodcarvings, though not as famous as those of their neighbors to the east, the Asmat, are considered by experts to be among the finest "primitive" art produced.

Until well into the 19th century, Europeans only sailed along this stretch of Irian's south coast without landing. The first useful map of the coastline was not drawn until 1905, when the Royal Netherlands

Geographical Society under A.J. Kroesen explored the lowest reaches of the Mimika River in a steam-powered sloop and made contact with the population. During the first decade of this century, several expeditions tried to cross Mimika territory to reach the snow-covered mountains in the vicinity of Puncak Jaya, which at the time was called the Carstensz Toppen, but the difficult conditions and the logistical problems brought on by the sheer size of these expeditions—hundreds of porters, tons of crated supplies—thwarted their success.

The first permanent Dutch outpost along the coast was established in 1926 at Kokonau, at the mouth of the Mimika River, and two years later, the Roman Catholic Church started proselytizing from the outpost. At first, the Mimika saw the church and western civilization as a godsend, as at the time they were suffering greatly from large-scale head-hunting raids by the neighboring Asmat, whom the Mimika called Wemanawe, the "maneaters."

In 1928, an Asmat war party of more than 100 men descended on the village of Atuka, on the coast 25 kilometers west of Amamapare. The Atukans who were not killed were chased into the forest, and the Asmat stripped the village, even taking the iron nails from the benches. The raid angered the Dutch police, who searched out and burned Asmat bivouacs and cached canoes. In 1930, seeking revenge, the Asmat returned with a massive force of, by some accounts, as many as 400 men, looting and causing great destruction to Mimika villages. When they tried again in 1931, however, the Mimika and the police had advance warning, and the Asmat were trapped between three-villages worth of well-armed Mimika and the Dutch police, the latter armed with guns. All but 16 of the Asmat were killed. The survivors were sent to jail in Fakfak. This permanently ended large scale raids by the Asmat.

At the same time the Dutch colonial government protected them from the Asmat, officials also forced the Mimika, traditionally a semi-nomadic people, into permanent settlements along the coasts for easier administration. The children, required to attend school, were abandoned by their parents for weeks at a time when they went upriver to process sago palm starch, and many of these hungry children succumbed to malaria and malnutrition. The Dutch tried to introduce a cash economy based on rubber, tea, coffee, coconuts and garden crops, but these crops did poorly, and the Mimika preferred their traditional lifestyle to that of permanent farmers. In the course of fishing, working the sago trees and hunting wild pigs and cassowaries, the Mimika moved around a lot, and enjoyed plenty of spare time to drink *segero* palm wine, carve, and practice the rituals associated with their traditional religion.

Dutch colonial officials strongly disapproved of traditional Mimikan religion, and eventually burned the ceremonial houses in an effort to "reform" the Mimika. Those cultural elements that raised the ire of the Dutch first went underground, then slowly died out. When the Japanese swept through the Dutch East Indies during World War II, the Mimika area marked the furthest advance of the Imperial Army along Irian Jaya's south coast. About 100 soldiers were stationed in the area, and they forced the Mimika to feed and house them, and to build an airstrip at the coastal village of Timuka (the name, altered slightly, was later transferred inland to present-day Timika). Anyone who refused was tied down to drown with the incoming tide, or simply hung.

Things did not much improve after the war. Mimika, stripped of its tradition by the Dutch, had become a dispirited, declining culture. In the late 1960s, Father Frank Trenkenschuh, of the Crosier Mission in the area, writes: "Mimika strikes a person as a dead area filled with zombies. There is no work and no interest in work. Religion of the past is no longer celebrated and the Christian religion means nothing to the people. The past is gone forever. The present lacks vitality. The future holds no hope." This grim description was penned about the time Freeport was laying plans to develop Ertsberg.

Contact between the Amungme and the outside world came much later than it did for their coastal neighbors. Reports of the 1909–11 British Ornithologists' Union Expedition to the snow-covered mountains speak at great length about a people they call the "Tapiro pygmies," who were Ekagi, the Amungme's neighbors to the west. The rather grandiose term "pygmies," which appears in the titles of two books about the expedition, was more dramatic than real—although most of Irian's highland people are relatively short-statured compared to the coastal Irianese or Europeans, they are not pygmies. To their frustration, the British were unable to obtain much information about these highland people, who, for very good reasons, were most wary of outsiders, including the coastal Irianese.

By 1936, when the Colijn expedition arrived, the Amungme were far less guarded. During the intervening two decades, they had no

Men inside a karapao, a Kamoro longhouse, in Tiwaka. Kamoro culture declined under the Dutch, who discouraged tradition religion and celebrations.

doubt heard of the Dutch government post at Kokonau, and some had perhaps even taken a look for themselves. When Jacques Dozy and Dr. Anton Colijn arrived in the Wa Valley, there was tension in the air, but when Colijn burst into song, it resolved into laughter. "We only saw one piece of iron, a relic of an old knife," Dozy said. "These people really lived in the Stone Age." The aim of the Colijn expedition was mountain-climbing and geological and geographical surveying, and the men did not have the time or skills to study the culture and social structure of the Amungme. But the next contacts by the outside world were of a very different kind.

Even before World War II, the Roman Catholic Church, from its original mission at Kokonau, had spread its faith among the Mimika. By the early 1950s, the Dutch Catholics were establishing a presence in the central and eastern parts of the Amungme lands. In 1953, the Catholic priest Michael Kamerer visited Wa, and the same year a Catholic delegation stopped at Tsinga. These first contacts by the church resulted two years later in mission outposts in the Tsing and Numa Valleys (the Numa is a highland tributary of the Cemara).

The same year, 1955, the American Don Gibbons, an evangelical Protestant with the Christian and Missionary Alliance, trekked over the mountains from his base at Beoga among the Damal, whose language closely resembles Amungme. Gibbons came to Beoga in 1956, built an airstrip, and remained there until 1994. Following a trail from the north side of the mountain range, he passed through the Wa Valley and spent the night in Opitawak hamlet on his way to Tsinga. The American missionary was looking for a place to build an airstrip to facilitate missionary work, but found no suitable site in the rugged Amungme lands. Also discouraging was the "significant Roman Catholic presence and a very strong, positive response to the Roman Catholic Church" he found in Tsinga.

Both the Dutch administration and the Catholic church found the Amungme lands too remote for their purposes, and sought to convince the Amungme to move closer to the lowlands. The site chosen was Akimuga, well down from the highlands in Mimika territory, about 85 kilometers due east of where Timika is today. The Dutch authorities had picked this area as a likely one for growing rubber.

This migration took place under the leadership of "Moses" Tembak Kilangin, an Amungme from the Tsing Valley who was educated in the Kei Islands and Kokonao, could speak four languages, and who had worked with Father Kamerer since 1953. In 1960 Kilangin and "Paulus" Zaelingki Tsolme, another Catholic village leader from the Tsinga area, led the first group of 30 Amungme from the Tsing Valley to Belakmakema, in the lowlands on the way to Akimuga, to prepare a way station for migrants. The first 50 families from Tsinga reached Akimuga in 1962; by 1970, 2,750 Amungme from Tsinga and Numa had migrated to Akimuga.

This migration was far from a success. The Amungme were unaccustomed to and weakened by the heat, and malaria—unknown in the high mountain valleys—was a recurrent problem. For example, the group of Wa Valley Amungme who followed Kilangin and Tsolme turned back at Belakmakema, finding the heat and disease intolerable before they had even finished half the trip.

When Forbes Wilson trekked inland in 1960 to collect surface samples of the Ertsberg, his contacts with the Amungme were quite amiable: "I was invariably impressed with their friendly nature and open generosity," he has written. But when Freeport began mining operations in the area in the late 1960s and early 1970s, tensions and misunderstandings grew between the Amungme and the outsiders.

The sudden appearance of 20th-century goods and machinery and western cultural values in the highlands of Irian Jaya shocked and confused the Amungme. Amungme society, and Melanesian society in general, is based a principle of equivalence in material goods. If a man has more pigs, or heirloom objects, than another man, these are given away as potlatch, which soothes jealousies and bolsters the giver's social standing. This flow of goods ensures the rough equivalence of wealth among all families while still creating charismatic leaders, or "Big Men" as they are called in Melanesia.

Freeport's remarkable objects—plastic raincoats, radios, bulldozers, helicopters—were not, as the Amungme expected, shared this way. The Amungme had shared their land, and this gesture was not reciprocated. Complicating this was the phenomenon of "cargo cults" widespread in Melanesia. In these "cults," traditional cultures, unable to comprehend the source of the very desirable material goods brought by outsiders, ascribe spiritual intervention to the production and distribution of these items. What the Melanesians see is that the outsiders have the ritual knowledge of how to obtain the goods, yet they selfishly keep this to themselves.

Cults based on the sudden appearance of western goods—

"cargo" in pidgin English, the lingua franca of most of Melanesia, means "goods"—have accompanied Europeans wherever they have arrived in Melanesia. Although usually associated with Melanesian culture, similar cults have taken place in other parts of Oceania, Africa, North and South America, China, Burma, Indonesia, Siberia and, at times in the past, in Europe.

Cargo cults, when they appear, often involve the building of the "ritual" objects that bring the goods: radios of wood and bark, scale model airplanes hewn of local materials, loading docks. Quite often these cult movements have a millenarian cast, with a secret society accumulating around a "prophet," who has the divinely inspired knowledge of a coming millennium of wealth and plenty. In some areas, these movements have grown so strong that whole villages have killed their livestock, razed their houses and waited for a redemptive event that, unfortunately, never occurred.

In the mid-1970s, a small cargo cult developed in the Tsing Valley, just east of Tembagapura. Frank Nelson, a Freeport geologist who worked closely with the Amungme while running our mapping and drilling programs from 1973 to 1977, remembers starting to see people wearing keys around their necks, like a crucifix or amulet. "The first time I saw that, I asked, 'Do you own a padlock?'" Nelson said. "I thought it was just a practical way to keep a key without losing it. Then I started seeing these all around, among our workers, and among other people, both men and women. The key was apparently something you needed to open the boxes when the cargo came." After a couple of years, the movement died down and people stopped wearing the keys.

Some of the early difficulties were less the result of cargo cultism than simple misunderstandings of the local political situation. Nelson's first drilling camp was on the Flint Extension—later the site of the Gunung Bijih Timur deposits—and when the drills had tapped out, he packed up and moved a few kilometers south, near the head of the Nosola River. This camp, Nelson recalls, was torn down three times by local Amungme. "They were very gentlemanly about it," he said. "They would wait until nobody was around, and they didn't steal a thing, which was the damnedest thing, especially as there were all sorts of useful things around." The first sign of trouble was a ring of hex sticks, called *salibs*, around the camp.

"First we found the *salibs* blocking the camp one day. They circled

Before Freeport and the Indonesian government brought organized health care to the area, infant mortality among the Amungme reached 50 percent.

it, and you're not supposed to go through *salibs*. We knew that there were people, hiding in the rocks, watching to see what we would do. So we made a point of not being overly impressed with them. Actually our local helpers, who were from Wa, peed on them. It was getting late, and we didn't want to be messing around, so we stayed there and went out the next day as usual. When we came back we found the camp torn down."

Nothing had been damaged or stolen. They had taken everything out of all the tents—clothing, tools, etc.—and put it in the mess tent. Then all the other tents were dropped. Lastly a *parang*—the Indonesian machete—was stuck into the main tent frame. Although the camp was useless, and everyone had to sleep that night in the mess tent, all the clothing and tools had been protected from the rain. "They were obviously saying, 'This isn't a big, dramatic thing, but we want to talk to you,'" Nelson said.

The perpetrators turned out to be Amungme from the Tsing Valley, and the problem was this: Nelson's mapping crew consisted of a team of geologists, and 52 local helpers from the Wa Valley just west of Tembagapura. Because the team had now crossed a watershed—the Nosola drains into the Tsing River—their argument was that Freeport should now be hiring its helpers from Tsinga. Since these early days, Freeport's field geologists have become more sensitive to "local hiring," even though the proximity of villages in some exploration areas means that crews that have received some training have to be swapped far more often than is optimum.

The culture shock from Freeport's presence hit hardest in the Wa Valley, just four kilometers west of Tembagapura. It quickly became clear that there were certain advantages to be gained from the company's close proximity, particularly medical attention, and the ready availability of food and goods. For example, something as worthless to a Freeport worker in Tembagapura as a discarded tin can could be of great value to an Amungme villager.

But resentment was just as common. In Forbes Wilson's *Conquest of Copper Mountain*, he notes that many of the people of Wa felt Freeport had wrongfully appropriated their land without adequate compensation. "When I retired in 1974 we still had troubles," he writes. "We were trying to figure out what to do about the hex sticks and angry tribal chiefs."

Indonesia's constitution clearly states that mineral wealth

belongs to the nation as a whole. The 1960 Basic Agrarian Law reinforced the 1945 Constitution, claiming all natural resources as national wealth, to be exploited for the benefit of the whole of the Indonesian nation. Freeport's contract of work to extract copper from Irian Jaya is with the Republic of Indonesia, not the villagers of Wa. Compensating the villagers for the use of their traditional land in getting this copper out falls in a very murky area of Indonesian law. It is also difficult to identify just how far from the operation this compensation should extend: the immediate residents of the valley, their extended clan, all ethnically Amungme people, all the highland groups, or all Irianese.

IN INDONESIA, CLEAR legal rights apply only to land under current cultivation. Ownership claims weaken considerably for land lying fallow, or used for hunting and gathering. For most Irianese, who traditionally have used large areas of land for hunting, sago gathering, and other purposes than planting, this makes land ownership claims very hard to press. There is a place in Indonesian law where these customary, or adat, rights are to be taken into account, but it is quite easy for government officials to re-assign the use of these lands if they consider it in the national interest to do so. It is also the government that decides on fair compensation for these adat rights.

Should it be necessary for mine operations, Article 2 of Freeport's contract of work gives the company the right to resettle people living in the project area. This is subject to our providing reasonable compensation for any houses and permanent land improvements which it becomes necessary to destroy.

Although Freeport's rights and obligations under this contract of work are clear, and legally binding, at first nobody explained this to the people of Wa—although so far none of our installations has displaced previously constructed housing or improved land anywhere in our project area. Also, at first neither the government nor the company recognized the importance of the area around Ertsberg to the Amungme, both as a place looked at from afar with awe, and as a traditional, though seldom used, hunting ground.

When mining began in 1973, a series of protests made it clear that the company had a problem on its hands. After negotiations, Freeport signed an agreement in 1974 with the leaders of the Wa Valley community stating that the company would build schools, clin-

Freeport's contributions to Banti Village include sturdy new *rumah sehat*, or 'healthy houses,' as well as a bright red tank for drinking water.

ics, markets, and a shopping center, and assist the Wa Amungme with accommodations and employment.

In exchange, the document states, "The local people of the region are ready and willing to give their approval for mining to go ahead at Ertsberg and other locations including Tembagapura and surroundings and in the ground underneath these locations." Six chiefs affixed their thumb prints, and Freeport thought that its problems with the people of Wa were resolved.

As the word about Freeport spread through Irian Jaya, many highlanders made the trek to Tembagapura. A few came just to have a look, but many sought opportunities and brought their families. Offering only unskilled labor, few of these newcomers obtained employment with the company. But rather than returning home, they hung around, hoping for something to turn up. Many of these were related to the people of Wa through the far-flung highlands kinship structure. Shanty towns sprang up at the edge of Tembagapura, at so-called "Upper Wa," along the road above the town, and "New Wa," near the helipad. Many of the people living in these shanty towns for-

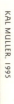

Freeport always involves residents when building housing. These in Nawaripi reflect the warm lowland climate, and Kamoro taste in materials and design.

aged from Tembagapura's garbage dumps. The shanty towns were a health risk and an eyesore, and the company was not happy with this state of affairs.

Negotiations took place to convince these new arrivals to either return home or to move down to the Timika area to more healthy and better managed quarters. This latter option was also offered to the original Wa Valley Amungme. Particularly in the early days, negotiations between our officials and the Amungme did not always go smoothly. The company was pragmatic, in its own way: we wanted to identify the leaders, negotiate with them, and be done with it. But the Amungme decision-making process is not so simple.

John Ellenberger is an evangelical Protestant missionary and anthropologist who lived for many decades among the Damal, who live just north of, and are closely related to the Amungme. In the early days, Freeport called him in on occasion to help sort out potential problems with the local people. Decisions among the Amungme, Ellenberger writes, "rather than being 'handed down' are 'discussed down,' resulting in a majority vote in which the minority by silence

305

gives consent to the decision." It is a process that takes considerable time, and oration by the Big Men is crucial. In essence, rather than make a decision for the people, the Big Men convince them.

In the mid-1970s, political troubles swept the Amungme of the Wa Valley and Tsinga. Following Papua New Guinea's independence in 1975, activities of the Free Papua Movement—Organisasi Papua Merdeka or OPM—a small guerrilla organization opposed to Indonesian rule of Irian Jaya, increased throughout the province. The area around the Freeport operations, which was considered a very visible symbol of Indonesian rule, was targeted for increased OPM organizing. In the spring of 1977, coinciding with Indonesian national elections, anti-government skirmishes occurred in various places in the highlands, and in July 1977, OPM protesters in the area around Freeport cut the slurry line with hack saws, felled a tree across a power line, and set fire to an oil storage tank. In the swift retaliation by the Indonesian military, most of the 17 new buildings Freeport had constructed for the people of Wa were destroyed, and some of the squatter camps were burned.

Following a flash flood some years ago, the people living in the shanty towns, which were vulnerable to flooding and landslides, were moved to the lowlands and the squats cleared. A new road has been built between Tembagapura and Banti, as the major village in the Wa Valley is now called, and a sturdy bridge and dozens of new houses have been constructed in Banti with Freeport's aid.

Health care has, in many ways, been the starting point of Freeport's many community development programs. "If you are borderline anemia all the time, you don't really have the energy or desire to go out and do much of anything different that you ever have done before," said John Cutts, who until moving to the New Orleans office in 1995 ran a number of community development programs for Freeport. In the highlands, the people are sapped by poor diet, and until the introduction of western medicines brought it under control, yaws or treponema infection. In the lowlands, malaria takes a constant toll. And everywhere in the tropics, relatively simple injuries, untreated, can lead to debilitating infections.

A new clinic has been built at Banti, with updated facilities. "This is a major clinic now," White says. "Right now we are treating an average of 130 patients a day. There are three nurses working here plus volunteers from the Tembagapura hospital." By mid-1996, the clinic will be converted into a "mini-hospital" with a 14-bed unit, and a resident doctor. It is mainly designed for convalescence under supervision, and serious cases will still be referred to the large, modern hospital five kilometers away in Tembagapura.

Freeport also built a hospital in Timika in 1994, which was subsequently turned over to the government. Recent expansions there, funded by Freeport, include an emergency room, an outpatient area, examining rooms, a record-keeping room, an X-ray machine and some laboratory equipment. Freeport built clinics at Aroanop, in the highlands west of Wa, at Kwamki Lama, and at Pomako. Once built, these were also turned over to the government to run.

The biggest health problem in the lowlands is malaria, which is widespread in Irian Jaya. In 1992, Freeport set up a large-scale malaria control program, run by Dr. Marvin Clark. Clark, a general surgeon who had practiced in rural Pennsylvania for 20 years, was first hired as a medical doctor in Tembagapura. He then began working on the public health committee and then the malaria control unit, which today operates with 89 employees and a yearly budget of $1.8 million. In 1992, fully 68 percent of the population contracted malaria; today this figure hovers between 2 and 8 percent, a percentage that compares favorably to other parts of Southeast Asia. The number of deaths directly attributable to malaria has, in the same period of time, dropped by 90 percent.

Case detection teams go out daily to all the populated areas of the lowlands, including the transmigration sites, the lumber camps and Timika town. The unit has calculated that over 80 percent of the entire lowlands population is regularly covered. For vector control, the unit maintains an on-going drainage program, with a team of 30–40 volunteers who dig and clear ditches. With rainfall in the Timika area averaging five meters a year, digging and keeping drainage ditches clear is a never-ending, but necessary task.

Fogging with insecticide is rarely done, as this tends to increase the resistance of the mosquitoes. Instead, the unit focuses on improving and maintaining drainage, and when necessary, larviciding—which kills the mosquitoes at the larval stage—using an environmentally friendly chemical called BTI.

During August 1995, a typical month, Freeport conducted 368 informal education sessions about malaria, with a total attendance of 9,247 people. All the transmigration areas, villages, hamlets, army

Though not formally a part of Freeport's recycling program, these children in Banti Village put cast-off inner tubes to their classic use.

posts and schools were covered in these sessions. During the same month, the unit supervised 72 drainage projects, manned by 1,617 community volunteers.

JUST LIKE BASIC health care, Freeport considers education crucial if the Amungme are to take advantage of the new opportunities now available in Irian Jaya. The company built a school in Banti in the early 1970s, but this building was one of the casualties of the troubles of 1977. The government was concerned that some of the people of the Wa Valley were implicated in OPM activities, and also faced a difficult logistical problem providing services in such a remote area. It wanted the village to move to Timika, and to encourage this move, the company was not allowed to rebuild the school.

Some of the Wa Valley Amungme did move to Kwamki Lama, but others resisted leaving their familiar cool, mountainous highlands for the hot, flat lowlands. After 15 years, with the political troubles having settled down and with the logistics improved, the government agreed to participate in a joint schooling program. Freeport helped

MAP 11.1

the villagers of Banti, Tsinga and Aroanop build "first primary schools" in their villages, offering instruction in the first three grades, and the government staffs the schools.

Of course, some of the children wish to continue with their studies after three years, and these must go to Timika. (The children of Freeport employees living in Banti can attend the school in nearby Tembagapura.) The company built a dormitory for the highland students, and currently 56 of them are finishing their elementary education at the Kwamki Lama school. Some of these will no doubt wish to continue through high school, and plans are being laid to accommodate their wishes.

In such a remote area as Irian Jaya, the availability of teachers is often a bigger problem than school facilities. Few of the well-trained teachers from western Indonesia consider a position in Irian Jaya as their ideal job. Because of this, Freeport has been providing teacher skill training within the province, sponsoring students to attend seminars at the University of Cenderawasih in Jayapura.

For students who wish to continue their education past the levels available in Timika, Freeport offers a scholarship program. About 30 scholarships a year are offered to send local students to the University of Cenderawasih, the largest in the province, and 250–300 more to send students to advanced high schools and universities both in the province and elsewhere in Indonesia.

One of the most successful educational programs has been vocational training. Timika's rapid growth over the last ten years provides a ready market for skilled workers. A job training program, started in 1990, currently supports 15 teachers and 160 students at the Vocational Training School in Jayapura. Closer to home, the company has just completed a Work Skills Development Center in Timika. Based on the company's pragmatic vision of providing basic, marketable skills to the local Amungme and Kamoro, the center offers training in skills such as carpentry, masonry, typing, bookkeeping, mechanics, welding and hotel work.

The school opened on March 1, 1996, with a dormitory for 60 students, housing for the teachers, and a classroom building. It is just west of Timika, and easily accessible by public transportation. Crucial to its success is that the teachers here are Amungme and Kamoro who are familiar with the language and learning methods of their ethnic groups.

Doug Learmont, who handles special training and environmental projects for us, planned the center. Learmont, who has been associated with the company since 1977, is married to an Indonesian woman, and has converted to Islam. He is extremely well versed in the language and culture of Indonesia. Freeport set up the Yayasan Irian Jaya in 1990, a community development foundation which Learmont coordinates. The Yayasan has played a role in most of the community development projects in the area, from clinic building and hospital expansions to Kamoro and Asmat woodcarving projects.

From the beginning, Freeport has attempted to draw local Amungme and Kamoro into the economic opportunities provided by the company's presence. Programs in Kwamki Lama, the Wa Valley, and other areas have helped to diversify and improve animal husbandry, fishing and agriculture. With the Kamoro, the emphasis has been on fishing and sago production; with the Amungme, growing vegetables. Freeport's hungry workers provide a guaranteed market for surpluses in both the highland areas and the lowlands.

In 1991, Freeport conceived an economic development program with a distinctly American flavor: the Business Incubators. This program aims at fostering entrepreneurship in the area by sponsoring small enterprises that provide useful goods and services to Freeport as well as other local businesses.

According to Kresno Wiyoso, who manages the Business Incubator Program, locally run businesses usually failed to compete with those run by outsiders from western Indonesia. "The Irianese level of education was quite low, and they had no access to capital or to markets, nor did they have the motivation. They had to learn about management, finances, product marketing, and consumer satisfaction, and how to handle employees and arrange production." The first group, Wiyoso says, took four years to "graduate," that is, until they could run their businesses independently. Now that the program is proven, the schedule is to graduate new entrepreneurs in two years. After four years, the program can be considered a success.

"The Irianese are realizing that going into business can be a viable option for them," Wiyoso says. "There are even some Freeport employees who are thinking of quitting work for the company to go into business for themselves."

A ready market is the most critical aspect of the program's success. Freeport remains the principal market, but Timika is now opening up and growing as well. "The company still has to import thousands of different items, more than 60 percent from outside Indonesia," said Wiyoso. "We can provide some of these items, with assured quality and at lower cost."

Yohannes Cenawatme, a 27-year-old Amungme, is set to graduate from the Business Incubator program. He is a junior high school graduate, and learned basic carpentry during a six-month stint at trade school in Jayapura. At the incubator, his carpentry shop churns out tamping poles, used to place explosives in blast holes, and

Freeport has helped build schools in Banti, Tsinga and Aroanop, and dormitories in Timika. For advanced students, the company offers a college scholarship program.

wooden survey stakes. He has recently diversified his product line to include core trays, and very solid ironwood benches for Kuala Kencana.

"I started off by myself, then hired four of my fellow Amungme," Cenawatme said. "As my business picked up I needed more skilled workers. Now I have sixteen employees, including some from Biak and the Kei Islands." Some of Cenawatme's carpenters earn up to $230 a month, which is a very good salary in Timika. Cenawatme himself has become solidly middle class. He has money in the bank, and sends his children to school. He has also acquired a second wife, a luxury that in retrospect may have been a mistake. Jealous quarreling between the two women is currently his greatest source of strife.

By 1995, twenty-one businesses had been created through the program. Six of these are run by Amungme, six by Kamoro, seven by Irianese from elsewhere in the province, and two are run by people from other Indonesian islands. Five of the businesses are run by women. To start these enterprises, Freeport provides the basics: capital and loan guarantees, an affordable space to do the work, shared support services, long-term purchase orders, and technical assis-

This ironwood park bench, destined for the new town of Kuala Kencana, is a product of Freeport's business incubator work skills program.

tance, including training in accounting and management, and help in obtaining commercial licenses and conforming to tax laws.

The two concrete block makers are thriving, thanks to all the construction in Timika and Kuala Kencana, as are the several nursery businesses, which produce and sell landscape plants, organic fertilizer, and orchids and cut flowers. Cloth goods businesses produce sample bags for Freeport's exploration program, as well as bed sheets, Indonesian flags, curtains, school uniforms and mosquito nets. Carpentry shops produce pallets and benches. A furniture shop produces rattan furniture. Two of the entrepreneurs run stores, one at the airport selling souvenirs and packaged food, and another at the Sheraton Inn Timika, selling Kamoro and Asmat woodcarvings. Ten local woodcarvers supply the Sheraton shop.

More than 200 local jobs have been created by these 21 businesses. They are still small enterprises, with assets averaging $10,000–$20,000, although one of the nurseries is a $180,000 concern. Since they are just starting out, the owners usually take only a very modest salary, about $150 a month, investing the rest in their business.

AMONG THE FIRST problems Freeport faced in its community development efforts were the often unrealistic expectations of the local people. John Cutts was raised by a missionary family in the Moni area north of the mountain range from the mine. He speaks Moni, and is well-acquainted with Papuan culture. When he first came to work for the community development department, he says, the Amungme would come to him with some unusual requests.

"The chiefs would come up to me and say, 'John, we'll need the helicopter for a day next week. We've got some people over the mountains with whom we've dealt in pigs and women in the past, and we'd like to fly over there with some pigs and shells, make a deal, and then come back. Can you please organize it for us?' In their minds they see this helicopter going everywhere and there's no reason in the world why they as the chiefs shouldn't be able to make pig deals with the Freeport helicopter."

The logic of this, at least from the Amungme perspective, is unassailable. And getting the Amungme to see things in a "realistic, mature way," Cutts says, is very difficult because they don't have any historical experience of the modern world. Manufactured goods, wage labor, even money were unknown to this culture until the late 1960s. "Some of their expectation levels go off the charts," Cutts says. "You get a bulldozer you can do anything. You get a helicopter you can do anything. You get a big truck you can load anything. 'Why make a big deal out of this, John? There are ships coming in with all this stuff, so what's the problem if we ask for a few things?'"

The easiest solution, of course, is simply to give the things away. The Indonesian government makes specific infrastructure requests and non-governmental organizations look for visible, measurable results—and fast. With the talent and resources we have on hand, it is no major accomplishment to build a clinic, or a row of modern houses. Send in a team of carpenters and a load of material, and they can be finished and back working at the mine in a day or two. But these kinds of gestures, while they look good in photographs and reports, do not often lead to *sustainable* development.

Stan Batey was hired by the community development department in 1992. He is a no-nonsense, salty-tongued Brit who has lived and worked in New Guinea for the past eleven years, first teaching agricultural techniques and animal husbandry on the Papua New Guinea side, then starting a successful wool-producing collective in Wamena, the largest town in the highlands, and the center of the Dani area of Irian Jaya.

"When Freeport got into community development, the local people thought that we should *provide*, since we were the big mining company, the Big Daddy," Batey says. "It was very difficult to change that perception, and it is still here today. What's more, we couldn't start at the grassroots level, because being a big mining company, people wanted to see some high tech." A middle road, which both meets the needs of the public relations image of a large mining company and is adaptable to the local people, is the direction the department tries to take. Balancing these often contradictory needs, Batey says, "is like being a gymnast on a four-inch beam. You got a bag of cement in one hand and a bag of cement in another and can't let either one of the bastards drop!"

For the past three years, Batey has helped organize the Community Project Center in Kwamki Lama. The center is concerned mostly with improving agriculture in the region, teaching the people to move away from traditional slash-and-burn methods, which are inefficient and damage the land, and teaching them how to husband new types of livestock: chickens, ducks, goats and cattle. The center's most successful group is the Sumber Jaya, a cooperative of about sixty farmers. "They are successful," Batey says, "because *they* came to *us*." The result was just the opposite for a scheme to get a group of Amungme women to farm peanuts as a cash crop. It failed miserably. "This idea came from our community development department and not from the grassroots level, the people themselves," Batey said.

In 1995, Agus Sumole took a leave from his professorship at the Manokwari campus of Cenderawasih University to take over as our manager of community development projects. Sumole is Torajan, from the island of Sulawesi, but he was raised in Irian Jaya. He holds a doctorate in agricultural anthropology from Queensland University in Australia. He is uniquely qualified for the job, and his personal philosophy meshes well with the current thinking at Freeport: The local people must be involved in both choosing development programs and in carrying them out.

"With the Amungme living near Tembagapura, we really have a problem," Sumole says. "Here we're talking about a generation already—twenty-five years—where these people have been exposed to Freeport just *giving*. We want to reduce as much as possible this

Freeport buys fresh labu siam for the mess halls from an agriculture project near Banti. Small-scale, locally initiated programs are often the most successful.

handing out of things. Of course, we understand that there are conditions where the people cannot provide for themselves. In the clinic, for instance, we cannot expect them to provide the medicines. But through education we do want to try to involve them more in the field of health, that the clinic should not be responsible for their health but they themselves.

"At the moment, I see our main role as preparing the human resources of the Amungme and Kamoro people," Sumole says. "We really have to start using the approach of developing what they have. For example, we tried to introduce agriculture among the Kamoro in Pomako. It didn't work. So what we're trying now is working with what they have, which is fishing and sago flour production."

The kinds of projects that make the most difference in people's lives are often simple, and receive little publicity. At any given time, Freeport is participating in dozens of these in both the highlands and the lowlands. Cutts recalls his first visit to Aroanop. "We asked them: 'What do you need? What would give us an opportunity to work together as a group, as a community, toward something that would really be of benefit to you?' And they looked around and said: 'You know, one of the things we really need is a bridge across this river, the Otomona.'"

The bridge across this river had been built, and rebuilt, many times. Rattan was simply not up to the rigors of the irregular, but inevitable flooding. When the bridge is out, it becomes very difficult for the villagers to get to their sweet potato gardens, which are on the other side from the village. Also, it is very easy to slip and fall from a rope bridge, and when this happens it not only brings grief to the family, but can also cause serious political problems. Within this highlands culture, if somebody falls from a bridge and dies, the blame for the death lies with the village to which the person was headed. In the past an event like this, which might be considered a simple tragedy in the United States, has been known to lead to warfare.

"That bridge turned out to be a really good project," Cutts said. "These people worked unbelievably hard to bring these huge ironwood logs out of the jungle, drag them down and plant them on both sides. It was a tremendous amount of work on their part, but they were very proud of what they accomplished."

Freeport contributed the time of several of its workers, Irianese trained in cable bridge construction, and provided steel cable, nails, some of the bolts and wire mesh to keep people from falling over the side. The material and the technical assistants were flown in. The total cost to the company for this project was a mere $16,000, but the value to the people of Aroanop is unmeasurable. Needless to say, the bridge is still standing.

Sometimes our employees have taken their own initiative in helping out in the community. Several years ago Wayne Cook, one of our best engineering problem-solvers, was running the tramway operation above the mill. A missionary in the Baliem Valley sent a message to the manager saying the people needed help building some bridges there. The manager passed the note to Cook.

"The upshot is there are three bridges I designed hanging across rivers over on the other side of the mountains," Cook says. "I went up, took a look, and measured things out. I did all the design in my spare time, Sundays and nights." By the time the two bridges at Mulia, and the one near Pyramid were done, Cook had spent four of his mid-year vacations supervising their construction. "It was something to do, something different," he says. "I took my family over there, and my kids had a great time, running around, playing with the Irianese kids."

BANTI IS THE closest Amungme village to Tembagapura, and this village today is radically different from what it was just ten years ago. From Tembagapura, a six-kilometer road leads to the Aghawagong River, where it meets a very solid suspension bridge. Over the bridge, a bright sign welcomes visitors to Banti. A wide road leads to the village office, and just beyond stands the large clinic, a complex of three buildings with distinctive hexagonal roofs. Next to the clinic is a three-room schoolhouse. Next to the schoolhouse are the new houses for the village head, and the six clan leaders of the area. Just slightly downhill from here, 46 sturdy two-bedroom houses with clapboard siding stand in two neat rows. Twenty-six of these are occupied by local Amungme working for Freeport. Clean water from a large holding tank is piped to communal faucets along the entire length of the village. (See MAP 11.1, page 309.)

Pigs, always underfoot in any other highlands village, are conspicuous by their absence. In Banti, the pigs have been exiled to a series of huts just outside the village. The owners feed them in the morning, release them during the day to forage in the hills for themselves, then tie them up in the late afternoon when the animals return for their supper. The village's sanitation is greatly improved by segregating these animals.

Creating Banti Village as it now stands is the result of years of patient negotiations between the Amungme and the Freeport development department. Just settling on the design of the houses required nine months of talks. But the villagers are very happy with the clinic, school, clean water and the bridge. Not everyone is convinced by the new houses, however, and most still live in their traditional round, wood-and-thatch huts.

While these traditional huts look picturesque to visitors, they are not particularly healthy places to live. Each has a fire going in the middle, and the ventilation is rarely very effective. Although they are warm and dry, the constant smoke creates serious respiratory problems for their inhabitants.

The Amungme recognize the advantages of these *rumah sehat*, or "healthy houses," in which the smoky kitchen fire is separate from the living space. But tradition is also strong. White recalls one of the Amungme leaders who, though he has installed his family in the new house built with Freeport's help, still lives in a traditional hut in one of the upper hamlets: "I'm too old, I like having my house up there," he said. "I'm glad my family is in the *rumah sehat*, but I'm staying in Utekini."

Freeport has helped build villages in the lowlands as well. Pomako, a small village on the Wania River south of Timika, was built—at the government's request—by Freeport. The village consists of 100 houses, a clinic, and a school. The people of Pomako are Kamoro who had been living in stilted shacks on Karaka ("Crab") Island just off Amamapare. Karaka had no fresh water, sanitary facilities consisted of the tide, and it was horribly overcrowded.

The 50 houses at Nawaripi, just south of Timika, and the 80 houses, school, church, meeting house and wells at Iwaka, twenty kilometers west of Timika, were also built with Freeport's assistance. Nawaripi Village was created to allow coastal Kamoro to live within ready reach of clinics, schools and economic opportunities. The improvements at Iwaka were in partial exchange for the land used for Kuala Kencana construction, some of which had served as hunting grounds for the Kamoro of Iwaka.

Not all of Freeport's material aid has been local. At Agats, almost

Three shy vegetable vendors at the Kwamki Lama market. Although traditionally highlanders, the Amungme have taken advantage of the opportunities in Timika.

200 kilometers southwest of Timika in the Asmat lands, the company has built dormitories and teachers quarters for a senior high school. In the same town, Freeport has built an extension to the Asmat Museum of Art and Progress. This museum, started by the Catholic Church, is unique among the world's museums of traditional art by not being in a North American or Western European capital. Instead its fine collection remains where the Asmat carvers themselves can see and appreciate their traditions. The light streaming into this new wing displays to advantage the stunning, large-scale sculptures of these famous artists.

CHAPTER TWELVE

Timika and a New Town

The empty jungle—An airport—Spontaneous settlements—Kwamki—A booming area—The transmigration program—Kuala Kencana—A model Indonesian town

IN 1967, THE area where Timika Town stands today was nothing but green jungle. Where today there is a bustling town of 35,000 people, then there was not so much as a small village. Taking off from Timika's airport, one now has to fly twenty miles east or west of the town to see empty jungle again.

The southern coastal plain of Irian Jaya was formed just 50,000 years ago, during the last great ice age advance, as glaciers tore up debris from the mountains, and rain sent it down the rivers. In the area south of the Sudirman Range, glacieologists believe that since this time the coastline has advanced as much as fifty kilometers to reach its current position. As it was formed largely from the decomposition of limestone rock, the soil was poor. The area was hot, swampy and malarial.

The Irianese never settled in this no-man's land between the mountains and the coast. They certainly passed through, on hunting, trading or war expeditions, but never settled. The climate and living conditions were—and still are—much better on the coast or in the highlands. Even today, outside of Timika, one can fly for what seems like hundreds of kilometers over the alluvial plain parallel to the coast without seeing any evidence of settlement.

Timika was born in the late 1960s, when Freeport decided to proceed with the development of the Ertsberg mine. The base of operations for the diamond drilling to establish the extent of the Ertsberg deposit was a small Kamoro coastal village called Timuka (marked "Timoeka" on the old Dutch maps). A more suitable and permanent logistics base was needed to develop the deposit into a working mine, and when a site was chosen—some 40 kilometers inland on the

Aikwa River—the name, in slightly modified form, came along. A small village still exists at Timuka, about 25 kilometers northwest along the coast from the mouth of the Tipuka River, and it is now called Timika Pantai, "Timika Beach," or Timika Lama, "Old Timika."

By 1971 the airport and logistics center had been constructed, and starting in the early 1970s a town began slowly to grow around them. When Freeport started construction on the mine access road in 1969, Amungme laborers were hired. Within a year, according to a survey conducted at the time by Moses Kilangin, 250 Amungme were working for Freeport at Camp II and at Mile 50. Within two years, these workers and their families had established themselves just north of the airport. This area represents the southernmost edge of Amungme territory, as the highlanders consider a set of rapids on the Aikwa River, at about the level of the airport, to be the division between their traditional land and that of the coastal Kamoro.

The Amungme cleared the airport area and opened gardens nearby. They were quick to move in to any area cleared for industrial purposes and then abandoned, such as temporary storage areas and construction facilities. They don't like to see any cleared land go to waste, and until recently spontaneous gardens created by these farmers in the open area established around the airport to meet aviation regulations were a source of continual conflict.

By 1973, the settlement north of the airport was large enough to warrant a name, and the Amungme called it Kwamki, which means "bird of paradise." It is an apt name, as these magnificent birds are still common in the area around Timika. The community also got its first village head, Paulus Magal, in the same year, and Freeport helped by bulldozing a side road from the main company road into the little village. This is the road that today runs past the market, and to the area called Cenderawasih, which is the Indonesian word for "bird of paradise." Freeport also cleared some space for houses and gardens. Shortly after Kwamki was established, the government set up a post in the area, at Mapurujaya, about 17 kilometers south of the airport. Government officials dubbed Kwamki "Harapan," which means "hope" in Indonesian, and this remains its legal name. But the inhabitants and the surrounding community still call it Kwamki.

After the Ertsberg pit began producing in late 1972, the company began experiencing problems with highland people settling around Tembagapura, in search of jobs with Freeport. Most of these were

Dani, Ekagi and Moni—not Amungme. Few found work with the company, as almost all of them lacked basic education and skills. A squatter community grew up outside of Tembagapura, which was not only an eyesore, but also a health risk because of the lack of sanitary facilities. Excessive tree-cutting led to the danger of landslides. The government was also unhappy with the squatters, as their location made it impossible for it to provide basic services.

By the mid-1970s, the government and Freeport began working to convince the newcomers to move to Kwamki, where there was space for proper housing and facilities. The Amungme living in the Wa Valley near Tembagapura were also invited to move, and houses would be provided for both groups along Cenderawasih Street in Kwamki. A total of 50 houses were built, most of them occupied by Dani and Ekagi, although a few Amungme moved as well.

The Amungme were the first to settle in the Timika are, but soon they were joined by migrants from western Indonesia. Many of these were Bugis, a Muslim ethnic group from South Sulawesi, with a few Kei Islanders, most of whom were Catholics. The Bugis are the best known indigenous traders of Indonesia, seafarers whose graceful schooners reached Australia long before the Europeans. The Bugis ranged far from their home in the center of the archipelago, seeking sea cucumbers, pearls and shells which they traded to the Chinese.

It wasn't until the late 1970s that a few houses began to spring up south of the airport, in the area we now call Timika town. This settlement was spontaneous. A house would spring up and from time to time, and suddenly there would be new residents. These first Timikans were a very mobile group, and few records were kept. Over time the same structure was probably occupied by a series of families as people moved through the area. At first the lowland was an unruly, even unsafe place to be. Then an army post was constructed near the airport, which brought a measure of safety and stability.

Additional houses were built in Kwamki, in several stages. In total, more than 200 homes were constructed. The focus at Kwamki was always to provide sound basic housing for the highland people, and to centralize facilities such as schools and hospitals. Providing such services in the highlands to the numerous small, semi-mobile Amungme settlements there was deemed too impractical.

During the political difficulties of 1977, the population of Kwamki, at the time mostly Catholic Amungme, fled to avoid the con-

Shopkeepers at the Timika market. Immigrants from western Indonesia moved to Timika beginning in the mid-1970s, taking advantage of the growing economy.

flict between the military and separatist guerrillas. They returned a year or two later, to find their houses mostly destroyed. These original Kwamki inhabitants then moved south of the airport, to an area that is now part of Timika, which became known as Kwamki Baru, "New Kwamki." The original settlement then became Kwamki Lama, "Old Kwamki," which it is still called today. The Catholic Amungme in Kwamki Baru enjoy close relations with the Catholic Kei Islanders who also live there.

In 1981 and 1982 the population of Timika grew further when the government relocated three coastal villages to the Timika area. The people from these villages were divided between an area called Koperapoka Baru, "New Koperapoka," named for one of the original Kamoro villages, and Sempan Barat, or "West Sempan," named for the Sempan, a Mimikan people who live along the coast east of the Inauga River, and whose culture and language are related to those of the Kamoro. Today these two districts within greater Timika are called Koperapoka and Inauga.

Many of the coastal people—unlike most Amungme—had been

A view of the dining area of the four-star Sheraton Inn Timika. This hotel, conveniently located near the airport, is probably the best in the province.

educated in Catholic-run schools and spoke Indonesian. The Kamoro and Sempan adjusted to life in Timika mostly by selling fish and sago, either on their own or through Bugis middlemen, who make most of the profits. The Bugis were the most dynamic group of the settlers. They set up shops and obtained construction contracts in the town of Timika from both Freeport and the government. The contractors hired local labor and paid very low wages.

In 1985, the company bought one of the Amungme garden areas to build Timika Indah, a development of houses for Freeport employees. This village was Freeport's first experiment in building housing to sell to its employees. At first we were uncertain that there was sufficient interest on the part of our employees to buy houses and resettle with their families in Timika. Much to our pleasure, we found a tremendous interest in home ownership. The 50 houses we built at first were not nearly enough. We could have sold four times that number. And the size and design of the houses we provided were deemed more than acceptable.

When the Indonesian government saw the developing population and infrastructure in the Timika area, it decided to set up official transmigration communities in the vicinity, and the first transmigrants—mostly from Java—arrived in 1985. Transmigration to the province of Irian Jaya began in 1970, and since then some 50,000 people have been relocated from the overcrowded islands of Western Indonesia. The population of Irian Jaya has barely reached 2 million people, while that of Java, little more than one-fourth the size of Irian, exceeds 120 million.

Indonesia's transmigration program has been controversial. It was started by the Dutch, who moved Javanese farmers to South Sumatra and Kalimantan, and was continued by independent Indonesia, originally with World Bank financing. Initially, transmigration sites were sometimes poorly selected. Frustrated farmers would try to grow rice or vegetables in poor soil, or they would succeed in getting out a good crop, and with no roads or other infrastructure, be unable to get their produce to market. Sometimes the land chosen for transmigration sites had been traditional hunting and gathering territory, which led to conflicts between the indigenous people and the transmigrants. Today, the program has been scaled back, and the officials in charge are much more careful about site selection and arranging a settlement with any traditional users of the land.

More than 2,300 transmigrant families—some 11,000 people—now live in the Timika area. They have been settled in seven numbered areas, each called an "SP," which stands for Satuan Pemukiman, or "settlement unit." The settlement areas around Timika are model examples of the program, and three of the SPs have already progressed to "desa," or village, status: Kamora Jaya, Timika Jaya and Karang Senang.

Freeport's purchasing power, the potential for employment the company offers, and the infrastructure it has created have directly or indirectly aided the transmigrants' success. Timika, easily accessible from the settlements, also helps by providing employment opportunities to supplement the transmigrants' income from farming, and a market for their produce. Because of the physical conditions and rainfall patterns on the south coast of Irian, very little irrigated rice is grown in the area, which is the specialty of the Javanese farmers. But many types of vegetables are grown, and the sale of these to Freeport and in town has provided the farmers with a decent income.

Given a chance, the transmigrants are a resourceful lot. They have set up small-scale businesses such as food stalls, soybean curd processing and various repair shops. They also work as taxi drivers, and jump at any employment opportunity provided by Freeport or its contractors. All this has led to the income of these transmigrants being among the highest of any similar group in Indonesia. One of the transmigrants puts it in more concrete terms:

"I miss my relatives back in Java, but when I was living there, I had a hard time earning two thousand rupiah [about $1] a day, the minimum I needed to eat. Here I earn much more than that, and even though everything is more expensive than in Java, I have enough to feed my family and even send my children to school."

Because of the economic activity of the new transmigrants—both government-sponsored and spontaneous—the town of Timika has boomed. By early 1996 greater Timika could claim 35,000 inhabitants, with some 15,000 concentrated in the urban area south of the airport. Central Timika includes two districts, Kwamki Baru in the north, and Koperapoka in the south, each with a population of more than 7,000 people. (See MAP 12.1, pages 328–329.)

The heart of the town is a large, covered market, Pasar Raya, formally called Pasar Swadaya Murni. This market, together with the adjacent taxi and minibus stands, the movie theater, the small shops, and the banks and offices, serves as the bustling commercial center of Timika. Space in this lively commercial center is at such a premium that today one is beginning to see three-story buildings being erected.

Timika looks like just about any other Indonesian town of its size, except for the almost total absence of ethnic Chinese–owned businesses, and that there are fewer street vendors. But the town offers all the basics: education through high school, telephone communication via credit card to anywhere in the world, post office, four hotels, including a four-star Sheraton, several movie theaters, three pharmacies, three banks, three mosques, four churches, and several restaurants. Light manufacturing has sprung up in the area, with several groups building furniture, and others producing concrete blocks, roof tiles and other building materials. Entrepreneurs in the area are attempting to cash in on a growing tourist trade, with shops specializing in artifacts and souvenirs from local Irianese craftsmen.

By 1993, with the potential of Grasberg established and a major expansion underway, the limitations of Tembagapura became clear. At this time, we began the process of establishing a major presence in

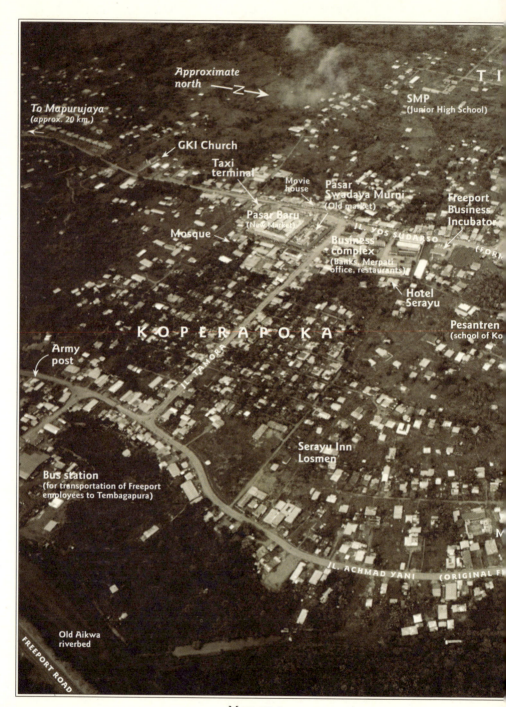

MAP 12.1

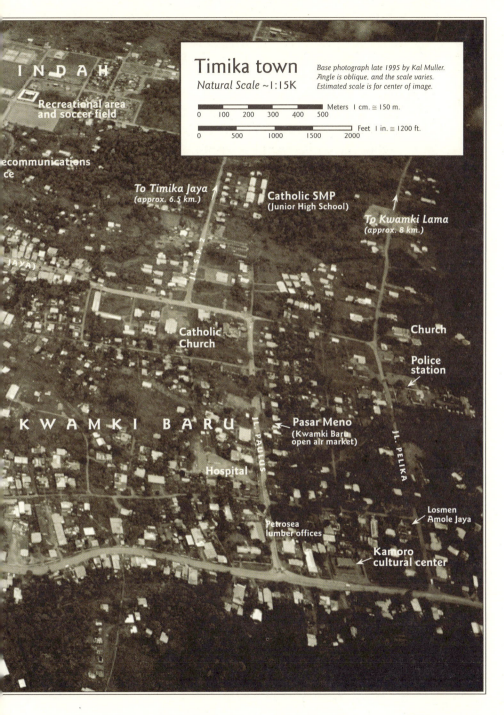

the lowlands. We decided to move many of our industrial facilities to the Timika area, and to continue with the program started a few years previous, of building housing and then selling the units to our employees, contractors and suppliers—or anyone else who needed them. We made the commitment that, over the long term, any function that did not need to be at Tembagapura would be moved to the lowlands, including the staffs and offices of any senior managers who did not need to be near the mine to operate efficiently. To accomplish this we planned a "New Town."

THE CONCEPT OF New Town grew quickly. It would be a model community designed to fit into its surroundings with a minimal impact on its environmental. While designs for the town were being worked out, we examined the Timika area for likely locations. Two areas seemed to meet our criteria very well. One was at Mile 50, along the access road, and the other was west of the road at about Mile 32. In 1993, both areas were unused, and Mile 50 at first seemed to be the most attractive, since it was 20 or 30 minutes closer to Tembagapura and the mine by road. Before we made a final commitment, we decided to test the soil for its strength in supporting foundations. Much to our surprise, Mile 50 had some of the weakest soils that we had seen in the entire lowland areas. To build on such land would be a costly effort, and we decided on Mile 32.

Before we formally committed to Mile 32, we once again checked for prior land use in this area. Although we could find no evidence of any previous use of this land, just when we began looking at it, a local timber company began pioneering a road through the area, felling trees on both sides. They claimed to possess the timber rights to the area, but they didn't have development rights to the land—we could build our town, they said, but we would have to wait until they cut down all the trees.

A town built on an unbearably hot, recently deforested plain was not part of our plan. We had visions of a model Indonesian community, sandwiched organically into the natural rain forest. We needed the rain forest—it was to be an integral part of our town.

We were able to stop the cutting, and then negotiations followed. To not cut the forest would be expensive for them, they said, as they had a plywood mill that needed the trees. As negotiations continued we could see this was going to be very, very expensive. In the end we purchased the timber rights for the area for $5 million, and were able to save this pristine area for the future town. We used their pioneer road as our access road, altering our design so we would not have to further disturb the forest by cutting another road through it.

We conducted a very careful topographic survey of the area, which had to be accurate to within one foot. Building in a rain forest is tricky. In order not to interfere with the natural drainage, and thus create stagnant pools where mosquitoes could breed, we had to survey and lay out the town very carefully.

The plan of the town was structured to fit into its natural setting, and at the same time, it was designed to be an Indonesian town, not an American town. Tembagapura is, in design, an American town that happens to be in the highlands of Irian Jaya. Our New Town would be an Indonesian town, through and through. Immediately adjacent to New Town we included an area which we dubbed the Light Industrial Park or LIP. This area would house light industrial facilities to service our mining enterprise, as well as those to service the construction and maintenance of New Town, such as sewer service, electric service and landscape nurseries.

The result of our efforts in 1996, with the first housing units constructed and most of the central facilities up and running, is, I believe, the best and most efficiently laid out small town in Southeast Asia. In 1995, as construction progressed with New Town, the concept and design began to achieve a high profile in Indonesia. We invited President Suharto to Irian Jaya to christen our new town, and to our delight, he accepted. Although we had to rush to get the final sewer, electrical and road work done, and to move in the first pioneer residents, it was worth it on December 10, 1995 when the president of Indonesia dubbed our New Town "Kuala Kencana." In English, the name means something like "River of Gold"—*kuala* is a river mouth, estuary, or confluence; *kencana* is a literary word for "gold."

Freeport old-timer Art d'Aquin, a veteran of all of our expansions since the 12K project, was a natural choice to oversee the construction of Kuala Kencana. He knew how the company worked, and knew what we needed to do to make our vision a reality.

"The primary purpose of Kuala Kencana is to give something new to our people, to create a better quality of life," D'Aquin says. "We have a great opportunity here; we are betting on the fact that we can find more ore bodies worth mining. This new town will become a

Indonesian President Suharto and Freeport CEO Jim Bob Moffett at the celebration inaugurating Kuala Kencana. Suharto is playing a Kamoro drum.

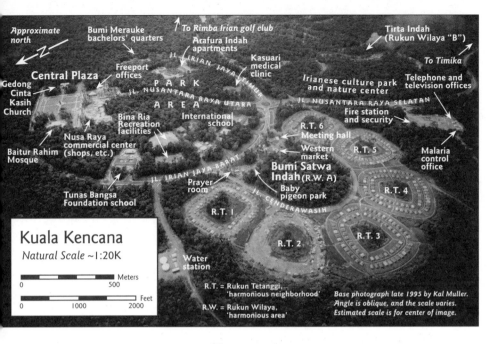

MAP 12.2

core, a pivot point for our operations. Tembagapura can't be that. It's in a place that just is not easy to get to. Kuala Kencana is built in a much, much better location. There's plenty of rain, but we have made many of the facilities all-weather—our people can play basketball or tennis or badminton indoors, year-round. We have over seventeen thousand hectares to work with, it's just a huge area."

The team in charge of designing the new town needed to know what its future inhabitants wanted. An American company was hired to survey our employees, and this information ended up in the Hay Report, named for the company that conducted the research. The researchers interviewed Indonesian staff and non-staff workers, and expatriates, and asked them a litany of questions: What don't you like about living in Tembagapura or Timika? What would you like to see done? What would make your life more enjoyable?

The Hay Report provided the guidelines. Then we selected a general architect to take these thoughts and comments, "Indonesianize" them and produce a master plan. The American team realized they knew next to nothing about designing and building for Indonesians,

333

who would constitute the majority of the future occupants of New Town. So we sought help from Bud Soehoed, a former member of the Freeport Indonesia board of directors, and a former minister and senior statesman in Indonesia. Soehoed now runs an engineering firm, and has been one of Freeport's wisest advisors for many years.

Soehoed's company conducted the initial field surveys, the soil, drainage and vegetation analyses, and some of the initial planning. Based on his company's work, we brought in a Spanish construction firm, Huarte, to do the actual building. Soehoed's work shows in the distinctively Indonesian center of the community, the *alun-alun*. This open, grassy plaza, dominated by a sculpture, is a pleasant contrast to a paved European plaza, enclosed by buildings. Nor did Soehoed's designers neglect the details. The kitchens, for example, have been designed to provide appropriate work space and storage for Indonesian cooks, who use almost exclusively fresh ingredients, in contrast to American cooks who rely more on frozen and packaged goods.

The Kuala Kencana square, the *alun-alun* of every self-respecting Javanese town, is the most visible sign that this is not Tembagapura. The size of this feature was a point of contention between our consultants. The Americans thought that 100 by 100 meters was plenty; Soehoed pushed for 300 by 300 meters. We finally compromised at 200 meters square. One might think that the axes of the *alun-alun* should correspond to the cardinal points of the compass, but here surveyor's logic took a back seat to religious imperative. The axis is tilted to square with the mosque, which faces Mecca in Saudi Arabia, the direction all Muslim worshippers face when praying. The classically designed Baitur Rahim mosque, with a capacity of 2,000, lies just off the northwest face of the square; off the southeast face is a large, multi-denominational church. To the southwest are a large retail complex and two Freeport office buildings. Although not yet constructed, eventually the northernmost face of the *alun-alun* will be graced with a government building, which will stand in the traditional position of the old *kratons*—palaces—of Java. The four sides of the *alun-alun*, taken together with the tall, central statue, represent five parts, symbolizing the Pancasila, the five principles which form the basis of the Indonesian nation.

Before any New Town construction began, botanists were brought in as consultants to identify and catalog the trees and other plant life of the area. In addition to the substantial green belts surrounding each group of housing units, every large tree that we could save in the construction, we worked around. Rain forest vegetation is basically very forgiving and the undergrowth grows back very quickly in small disturbed areas—but the large trees do not.

The first phase of Kuala Kencana, completed in mid-1996, has housed many of the employees not essential to the operations of the mine and the mill, including those in the accounting group, the human resources personnel, the government relations department and many of the administrative staff.

Until the construction of Kuala Kencana, many of the mine workers lived a bachelor's life, working hard and living in dormitories up near the mine, and visiting their wives and children back in Java, or Sumatra, or Sulawesi only during their scheduled home leave. The completion of the new town, together with a change in 1996 to 12-hour shifts, will make it possible for them to bring their families to Irian Jaya. A worker's family can now live in Kuala Kencana, and the miner can spend approximately half of each week with them, and the other half working the long shifts up at the mine.

Kuala Kencana has plenty of facilities for families. Already there is a 12-classroom foundation school for the Indonesian children, and a smaller international school for the children of expatriate workers (Indonesian law does not allow mixed classes of expatriate and national children). The first phase of building includes a medical clinic, although serious cases will still be sent to the Timika or Tembagapura hospitals. Eventually Kuala Kencana will have its own hospital. Extensive recreational facilities are already in place: an olympic-sized pool, soccer and softball fields, and badminton, basketball, tennis and squash courts, many of them covered for all-weather play. The retail and entertainment center includes a department store, grocery store, restaurants, two movie theaters and a bowling alley.

A town of the size of Kuala Kencana requires substantial infrastructure—power, water, and sewer systems. The 9.4 megawatt power station at the Light Industrial Park was upgraded by 4.7 megawatts. The malaria control teams drained all stagnant waters in the area, and remain vigilant in their ditch-clearing. The town's water supply comes from three 50-meter-deep wells, each pumping 25 liters a second. The sewage treatment plant, 3 kilometers west of town, is a state-of-the-art facility, and ensures that nothing harmful

The central alun-alun of Kuala Kencana, with the Baitur Rahim mosque. The large sculpture depicts four Kamoro shields and swirling birds of paradise.

Kuala Kencana, or 'River of Gold,' fits organically into the surrounding forest. In the foreground is Bumi Satwa Indah, the first neighborhood completed.

remains after the processing. Fire and security buildings, with live-in accommodations, are located right at the main entrance to the town.

One of the prominent and attractive features of the town is its championship golf course. Designed by Ben Crenshaw, who won the 1995 Masters Tournament, and Bill Coore, the 18-hole, par-72 Rimba Irian Golf Course winds for 6,850 yards through spectacular primary tropical rain forest. The course was constructed by Rod Whitman, in association with an Indonesian company, P.T. Kamanta. The 40-meter-wide fairways are lined with vertical walls of vegetation shooting up to 35 meters. The crowns of the largest trees reach nearly 50 meters. The more dramatic trees in the way of the course were accommodated as much as possible, and detours were constructed around several magnificent strangler figs. Because of the frequent downpours, an elaborate drainage system was constructed, requiring 110,000 cubic meters of sand to be placed under the turf, and above the alluvial gravel. This system is effective enough to allow play to resume within an hour or so of even the heaviest tropical rainstorm. The Rimba—the word means "Jungle"—clubhouse has striking

337

architecture, incorporating huge local hardwood beams, shaped by hand and hoisted into position with cranes. Everything except the roof, which is slate, was constructed of local materials.

Near Kuala Kencana, but separated by several kilometers of undisturbed jungle, is the environmentally sensitive Light Industrial Park. In addition to the power plant, the LIP includes a steel fabrication shop, a fueling station, an oxygen and acetylene plant, and food warehousing facilities. Originally conceived to handle industrial activities supporting our mining operations, the LIP has now been classified as a duty-free economic development zone. More than 50 companies have already shown interest in locating at this industrial park. These businesses will provide local employment opportunities, and the Freeport-sponsored business incubator program will insure that some of these—so far including home delivery of liquefied natural gas, landscape maintenance and carpentry work—will go to local Irianese.

The first 66 families who moved into Kuala Kencana found a neighborhood of pretty, well-designed houses waiting for them. The houses come in two sizes: 54 and 72 square meters. The town is divided into six large clusters of houses, called Rukun Wilayah, literally "harmonious areas." Only the first of these, R.W. A, or Bumi Satwa Indah, has been completed. Each R.W. is then further divided into six smaller units, called Rukun Tetangga, "harmonious neighborhoods." The houses in the Rukun Tetangga are arranged in rings, each surrounding a green area with a playground for the children. The six Rukun Tetangga are themselves arranged in a U to form the Rukun Wilaya. Satwa Indah includes a large green area, a meeting room, a supermarket, a kindergarten and children's park, and a prayer room.

Rukun Wilayah B, scheduled for occupation in mid-1996, is of a slightly different design. This development, the Tirta Indah, is designed for Freeport staff and will include larger houses, arranged in a more free-form pattern. It also has its own kindergarten, local community building and recreational facilities. In the inside of the central loop area are a smaller development of three-bedroom apartments, the Arafura Indah, and one of single-bedroom bachelor's quarters, the Merauke. Kuala Kencana's most coveted housing will be the large, 160–250-square-meter estate houses now being built next to the golf course.

THE FIRST FAMILIES moved to New Town in a bit of a rush, to get there before President Suharto made his inauguration. These were families of Freeport workers who were already living in Timika. While there were some grumbles about the lack of built-in air-conditioning—some people prefer fans, so this is a buyer-installed option—there was only one major oversight. The planners had forgotten to provide a covered area in back for drying clothes out of reach of the frequent rains, so the bathrooms had to double temporarily as laundry rooms.

Despite these small complaints, just a couple of weeks after the inauguration, the new residents of Kuala Kencana were amazed at the change in their lives. They now own their houses, and their mortgage payments come to considerably less than they were paying in Timika to rent houses that were not nearly as spacious or modern. The electricity is on 24-hours-a-day, unlike Timika where it is available only at night. The parks, stores, sports fields and other facilities are nearby and of high quality. Their children offered their own ringing endorsement of the new town—they were playing and running around as though they owned the place. And thanks to the well-planned layout, they have their run of a practically traffic-free street or a central playground, always close enough for Mom to keep an eye on them.

CHAPTER THIRTEEN

The Future

Bright prospects for growth—Exporting technology—A bridgehead—Skilled local work force—Expanding provincial infrastructure—An agricultural center—Continuing demand for copper

As we move into the 21st-century, the future looks bright for P.T. Freeport Indonesia. With the completion of the 115,000 tons per day expansion—which has already been reclassified as 125,000 tons per day—Freeport's operations are producing concentrate in a manner increasingly fine-tuned to minimize costs. A related expansion of Freeport's smelter in Spain has been finished, and the construction of a new smelter is soon to start at Gresik, near Surabaya on Java. Meanwhile exploration teams in and around the Grasberg deposit, as well as east and west of Grasberg along the central mountain range, are coming up with encouraging results.

In short, Freeport's prospects appear extremely good in the closing years of the century. There will no doubt be additional expansions of production facilities, as well as additions and improvements to our infrastructure. It is difficult to predict with any precision how things might unfold over a period as short as, say, the next five to ten years. So, in this chapter, I will take an author's liberty and speak of the trends I see developing in the long term—15 to 20 years and beyond. These are trends I see affecting the company, its stockholders, its employees, the surrounding community, the world of copper, and the government of Indonesia—a group that I would call Freeport Indonesia's "stake holders."

Freeport-McMoRan Copper & Gold, and its main production subsidiary, P.T. Freeport Indonesia, has chosen Indonesia and the island of Irian Jaya as its theater of operations. This decision was based upon the excellent opportunities for growth provided by the favorable geology of the region and, in particular, the astounding potential—much of it already in hand—of the area around Grasberg.

In the future, Freeport's mine operations will support a diverse and successful community who for three or four generations will have been associated with the mine.

Freeport Indonesia today has a major exploration program in the immediate area of Grasberg, as well as in remote areas of the province. As of this writing, more than 75 prospects have been identified. Although this sounds promising, I must caution that in this industry probably 100 prospects of this quality must be examined to find a single deposit capable of supporting a major mining operation. This process of finding and examining these prospects is expensive and time-consuming.

To work through the inventory of prospects we now have available would typically require five or more years. Once a major discovery is identified, feasibility studies must be conducted, marketing and financing arrangements must be negotiated, and finally the construction of a project must be undertaken. The process from the discovery of the resource through construction of a new mine operation typically takes 5–7 years.

Having said this, I am certain that within 15 years, Freeport Indonesia will be operating a major new base metal project remote from the existing arena of operations within Irian Jaya. The 75 prospects I just spoke of—together with others yet to be found—are sure to yield success. Also, I predict there will be at least one smaller but economically important high-grade gold operation. These new mines will be in addition to our existing operation, which will continue to be dominated by Grasberg.

In this same 15-year time period, additional reserves will be created from prospects we know of and others yet to be found. Even 15 years from today, the 100-square-kilometer COW around the old Ertsberg will still enclose 30–40 years of reserves. By then the Block A operation will have been expanded to 200,000 tons per day, or perhaps even more. There probably will be an entirely new concentrator included in this 200,000 TPD expansion, which will be designed to handle ores of a different nature than we see today.

The Grasberg porphyry will still dominate production, but some of the peripheral skarn deposits surrounding Grasberg will have begun to come into place, such as Lembah Tembaga, Kucing Liar, and Big Gossan—and others which will be discovered as the Grasberg operation evolves. In 15 years, underground mining will again be a major part of the mine production technology for the Grasberg area. The underground mining techniques will include block caving, sub-level caving, and long-hole open stoping. The company will have again successfully faced a series of major challenges in the introduction and reintroduction to Irian Jaya of these diverse mining methods.

This pace of growth will probably force PTFI to look abroad for opportunities. The scale of its operations will demand a major participation in overseas markets for its products in Asia, Europe, and North America. It may even be necessary to participate in downstream processing and manufacturing facilities to assure the movement and sale of its copper.

Further, within 15 years the technology developed and used by Freeport Indonesia may itself become a marketable product. Already, PTFI is a pioneer of technology at the leading edge of the copper industry. Developing and further refining this technology, together with the training of personnel at all levels, will create an opportunity for growth in technology production—an opportunity which perhaps should not be limited to Indonesia.

What does all this mean to the shareholder 15 years from today? The growth I foresee is a continuation of that we have seen in the past. The management of PTFI has always had the "delightful problem" of more opportunities than it has had the funds to develop. In today's world, opportunities are measured by the rate of return they deliver to the investor, and Freeport Indonesia has always had an abundance of projects available that would produce a return on capital of greater than 20 percent after taxes. This condition exists today, and I see no reason why it won't continue.

As the operation grows, our ability to take advantage of these opportunities will increase. Therefore I expect that we will see production, as measured by sales, double—and then double again—in the next 15 years. Where today PTFI produces 5 percent of the world's newly mined copper, in 15 years it could be producing 20 percent. This level of growth will be as rewarding to the stockholder as the company's past growth has been. Just like the last 15 years, growth over the next 15 years will not be even. The nature of the metals market produces cyclical prices, which follow the ups and downs of the world economy. Because it is such a low-cost producer, Freeport Indonesia will be able to weather the low periods better than other copper producers, and will be able to maintain a steady course of planned growth. This should result in excellent rewards for the stockholder, as well as for all the other stake holders.

THE CONTINUING HEALTH of Freeport's Indonesia operations will provide long-term growth opportunities for its employees and their children. We have around the Grasberg deposit an opportunity that is seen in few other mining areas around the world—an operation that will last more than a 100 years. This reminds me of my time in South America, where it was very common to meet people who were members of the third and fourth generations of their families that have worked for the same operation. I fully expect this will happen in Irian Jaya.

Again like in South America, the diverse group of people living and working in our mining district will all refer to themselves as natives of Irian Jaya, although their ancestors will have come from all parts of Indonesia. The majority of this group will have its family roots in Irian Jaya and eastern Indonesia, but Java and Sumatra will also be well represented. Before our mine operations began, there were at most a few hundred people living in the area. Today, not even 25 years later, with 50,000–60,000 people, it is one of the most populous parts of the province.

With the continuing success of its training programs, Freeport will someday not only be self-sufficient in providing its own labor force, but will also export skilled labor.

One hundred years from now, the group of people living around the Freeport Indonesia operations will be well-educated, maybe even world leaders in the art and science of the mining industry. Their home will be Kuala Kencana and its surroundings, including Timika and Tembagapura.

By necessity, this group will be familiar with a wide range of technologies, as they will staff the major industrial complex needed to service the mine operations. The managers in this group will be expert at both mining operations and international metals commerce.

Today, fewer than one in three of the people living in our project area have a direct association with Freeport. Still, it is Freeport that provided the incentive, the logistical means, and the economic capability for this group to function as it does today in Irian Jaya. I am sure this trend will continue. Any major industry produces a multiplier effect, and this can certainly be seen today around our operation. The cash generated by our mine sends ripples flowing outward—first through Irian Jaya, and then through the rest of Indonesia.

We are only now beginning to see the true scope and importance of the Freeport Indonesia project to the surrounding community. I believe the operation's size, and the engineering challenges it has faced, has obscured the most important aspect of our mining project. This is the bridgehead and logistical corridor that we have created from the Arafura Sea to the very crest of the central highlands. This is a very valuable asset, and it will shape the area for generations. I doubt the government would have been able to find the funds or justification for this endeavor for at least another half-century.

This bridgehead will eventually bring the benefits of modern civilization—and its problems—to the large group of people living in the area from the southern watershed of Irian Jaya to the central highlands. This is something the people of the region eagerly anticipate. They have already demonstrated this approval by voting with their feet—migrating to the area from all over the province.

This development will also bring the outside world to Irian Jaya, which also has its pluses and minuses. Already, a growing number of people come to the area to see, and participate in, the marvels there. After the start of the mine, perhaps in recognition of the eventual importance of this bridgehead, the large nearby Lorentz reserve was established. In time, there will be daily tours into this national park, both in the highlands and the lowlands.

The area offers additional opportunities for the development of land for farming, raising livestock, fisheries, mining, and industries of all types. In the beginning, perhaps, these will be driven by the availability of local resources. Among these is water, provided by the area's high rainfall. This abundance of water brings an opportunity for hydroelectric projects, thus providing the corridor with a clean, cheap source of power. My prediction is that these projects will not be gigantic dams, which require flooding large areas. More practical will be smaller, run-of-the-river, high-head projects of the type one sees today throughout Japan.

The boundaries of what can be considered the surrounding community will also change. As mining operations remote from Tembagapura are developed, the infrastructure necessary to these will expand the size of this community. Surely, one of these new operations will be north of the mountain range. Its location will probably force the construction of some form of access road from the north coast of Irian Jaya. This could result in a completely new port and logistics operation on the north coast. Eventually this will be linked to the south, creating a true trans-Irian road.

Here the government's plans to develop the area will dovetail neatly with Freeport Indonesia's plans. This road, in fact, is already in an advanced state of construction. The most difficult link, from the South Coast to the crest of the mountain range near the Grasberg mine, has already been completed by Freeport. The Nabire–Ilaga sector, currently under construction by the government, is nearing completion. Once finished, this road will bring the community along this corridor into the twentieth century, and it will join with the growing South Coast community. After the main north–south road is constructed, east–west branches will follow.

For PTFI, the completion of this route will improve the economic feasibility of mining operations in the area. A mine that may have offered a marginal return when factoring in the cost of constructing an entire access infrastructure, may become quite profitable if a nearby road already exists.

This road will bring an important change to the central highlands. This is where the greatest portion of the Irian population lives, yet even in today's modern world, this community is cut off from the outside world. Everything reaching the highlands today must come by air, which severely limits the scale of development that can take place.

The completion of this logistics and transportation link, started by Freeport in the late 1960s, will bring hundreds of thousands of Irianese highlanders into the modern age.

In Timika, seeds are being sown which I believe will result in a remarkable scene within fifteen years. Schools and industrial training centers are now being built to provide PTFI with a pool of people who have the work and supervisory skills we need. Freeport's needs outstrip the capability of all the training and educational facilities in Irian Jaya. The company is sometimes criticized for not hiring enough people from Irian, but we are also criticized for hiring too many—there are still so few skilled and trained Irianese that by hiring those we do, it is said, we take all the talent, thus retarding the development of the rest of the province.

Within a short time, Freeport will become largely self-sufficient in supplying its local personnel needs. In fact, our training and educational facilities may some day exceed our needs. In this case, the facilities in our area will begin to export these "resources" to other parts of Irian Jaya. It could well be that an institution like the University of Cenderawasih, based in the provincial capital of Jayapura on the North Coast, will—with Freeport's support and cooperation—establish a branch campus in our area for advanced education.

The facilities that a future Timika could support would easily exceed in size and scope those the university currently maintains in Jayapura. We may even see a College of Mineral Sciences churning out mining engineers, metallurgists and geologists for the mines of Irian Jaya and the rest of Indonesia.

The area, in time, will also support a major agricultural industry. Again, the trans-Irian highway system and other logistics facilities are what will make this possible. The highway will open up much currently vacant land, and it will pass through a variety of climactic zones, allowing a variety of crops. Beginning at the South Coast, the road will gain altitude as it approaches the highlands, then gradually drop off again on the north coast. In the cooler, higher elevations, cash crops such as coffee and tea could be grown, and in other suitable areas, various fruits. Feed lot and cattle ranching will also be possible. A local base of cash paying customers is already supporting a small feed lot industry. With the establishment of additional local sources of feed, this industry will grow, and there will soon be an opportunity to ship its product to customers outside the area.

In the future, the Freeport access road will likely be linked to a road from the north coast, creating a corridor of opportunity for the people of Irian Jaya.

The seacoast in southern Irian Jaya abounds with potential. Baramundi fish and mud crabs, common in the area, are two potential high-value products that sell well in the Australian and Japanese markets. The logistics base on the coast provides ready access to other Indonesian ports and overseas. Today, many of the cargo ships arriving in Irian Jaya leave empty. The incremental costs for these ships to carry agricultural cargoes away from Irian Jaya would be very small.

Kuala Kencana, with its Light Industrial Park, and the incubator projects around the Timika area, provide additional potential for growth. During Freeport's most recent expansion program, major heavy steel fabrication was undertaken in the LIP. This proved to be very competitive with steel fabrication in Java or overseas. Continuing expansions, as well as servicing PTFI's ongoing needs, will further hone the expertise of the personnel working in this area. A major steel fabrication industry could easily build up in the area.

The possibilities of tourism are also great. Timika already has all the basics. It has a four-star hotel that sits beside an international airport. Just down the road is a well-designed championship golf

The slurry lines from the Grasberg mine now carry 5 percent of the world's newly mined copper. Within 15 years, this could easily rise to 20 percent.

course, complete with a clubhouse. The course was designed especially for the frequent rains in this area, and the drainage is so good that it is possible to play this course just 60 to 90 minutes after a heavy rain.

Timika is in the same time zone as Tokyo. The day will come when Japanese tour groups will fly in for a week of golf and exotic touring. Golf will take up much of their time, but tours to Agats, the Lorentz Park, the central highlands, and the glaciers will be on the schedule as weather permits. Within 15 years, as travel by highway to the central highlands becomes possible, we will see this take place.

The surrounding community will grow, in time, to include other parts of Indonesia. Today, Freeport is about to embark on the construction of the first grass-roots smelter that the copper industry has seen in 15 years. This will be constructed at Gresik in the province of East Java. The smelter will be built next to a fertilizer plant. One of a copper smelter's major by-products is sulfuric acid, the raw material for manufacturing fertilizer. Today the fertilizer plant imports sulfur, and burns it to make sulfuric acid. When our smelter is complete, the fertilizer plant will buy its sulfuric acid from just across the fence.

Initially, the capacity of this smelter will far surpass the copper needs of all of Indonesia, and approximately 75 percent of its output will be exported. Over time, however, Indonesia's needs will grow—and less of the Gresik plant's output will be exported. The presence of a ready source of high quality copper in Java will eventually foster a healthy copper wire and electrical products industry. In this way, the ties between western Indonesia and the Grasberg project in Irian Jaya will strengthen.

One segment of the surrounding community deserves special mention—the Amungme. Fifteen years from now, this group will have come the furthest, while at the same time staying in the same place. Before the mine started, life for them was short, and its quality by any measure was substandard. Small tribal groups were in constant warfare with neighbors just over the ridge in the next valley.

In the early days, I remember frequently seeing children and adults in the area with reddish hair—a sign of malnutrition. These indigenous people have been very quick to see an opportunity for a better life. The leaders among them recognized that they did not have the technical skills to participate in the opportunities around them, and that these can come only with time and training. But they also see a great opportunity for their children.

Fifteen years from now, the scene will be very different. Large numbers of highland people will be entering the work force, and their education level will be equal to—or even better than—Indonesians from other parts of the archipelago. These Irianese will be familiar with mining and will be physically well-adapted to the climate and altitude. They will make very tough competition for newcomers, and will begin to dominate the work force.

WITHIN 20 YEARS, the government of Indonesia will realize most of the objectives it sought when it first made the decision to enter into a contract of work with a foreign company to develop Ertsberg. Spearheading the development of the region was clearly one of the government's first objectives, and this has already been achieved. The financial benefits to Indonesia are also considerable.

Today the government owns a bit less than 10 percent of Freeport Indonesia, a share that has a value of more than $400 million. In 20 years, with the continued growth in the area, this investment will pay off handsomely. I also believe that within this time, the government will not only reap the benefits of this growth, but will also liberalize its investment laws.

If this becomes the case, P.T. Freeport Indonesia, by means of Freeport McMoRan Copper & Gold, will be listed on the Jakarta stock exchange as well as the New York stock exchange. The government will convert its PTFI shares into FCX shares, and sell them on the Jakarta exchange—thus increasing the investment opportunities available to Indonesians and raising cash at the same time. The dual listing of the stock will allow the arbitrage of prices between these two exchanges, and will make the stock a point of interest for traders around the world.

The Indonesian government is also a recipient of tax and other economic rent from the PTFI project, and by 20 years from now, the income from this source should have ascended to the multi-billion dollar level.

These are only direct payments. A basic industry like copper mining spins off many other benefits—the multiplier effect mentioned above. The results from this can be seen in the increased level of commerce rippling through the economy.

Perhaps the greatest accomplishment for Indonesia as a whole of those in the government in 1967 who were the architects of the first contract of work is this: the fears of foreign investors have now vanished. In the late 1960s, the investment atmosphere was clouded by worries about nationalization, currency conversion problems, dividend restrictions, and import and export restrictions. Today, with the growth and success of the Freeport Indonesia project, these fears have been laid to rest.

Twenty years from today, Indonesia will continue to have a controlled, but well-managed, economy and government. At that time, I predict, the scene in Indonesia will very much resemble Japan in the 1960s. The new investor in Indonesia 15 years from now will face none of the worries we did in 1967, and Freeport, as the Suharto government's first foreign investor, will have led the way.

PRODUCING BETTER THAN 5 percent of the world's newly mined copper, P.T. Freeport Indonesia today is one of the copper industry's major players. With the growth over the next 15 years that I predict, PTFI's importance could well approach that of the Chilean

Amamapare in the future could be the center of shipping for the diverse raw materials and manufactured products moving in and out of a thriving economic zone.

state-owned copper mining company, CODELCO. The future and the well-being of copper, and its intensity of use in world economies, will become increasingly important over the next 15 years. Some of the events and developments taking place today can give us clues to where the world of copper is heading.

With the close of Cold War hostilities—and the resulting political turmoil and suspicion of investment that it brought—we are seeing stability, development, and growth in new parts of the world. There is now a huge potential for economic growth in countries like Brazil, China, India and Indonesia, places where there are large populations and undeveloped or underdeveloped natural resources.

The intensity of copper use in these countries is today quite low. But as they develop, and their standard of living increases, their copper consumption will grow, perhaps even reaching the levels now seen in the developed world. This factor has driven the copper market for last 2–3 years, and I expect it to continue.

If demand exceeds supply, high copper prices could act as a brake on the industry's growth. Copper is a commodity, and at higher

Dried concentrate being loaded in a cargo ship at Amamapare. Increasing demand for copper in the underdeveloped world will drive the market in the next 15–20 years.

prices, consumption drops. The usual result is that decreasing consumption brings a drop in the price of copper, and eventually production of copper tails off. This then brings an increase in demand, and higher prices, and the cycle repeats. The added production that will enter the market from Freeport Indonesia and others should act to attenuate this cycle, keeping prices reasonable. Even with the ups and downs of the copper market, when looked at over time in constant dollars, copper remains a bargain, and copper prices should not act as a serious brake on growth.

A more serious threat to the world of copper comes from environmentalists. Our lack of precise knowledge as to the role of copper metal in the environment leads to uncertainty, and perhaps unreasonable regulations. Just what level of copper should be allowed in the environment is still unknown. When does copper become a health problem? It is important to keep in mind that copper is essential for life—a lack of copper can result in death—and death from the intake of copper is almost unknown.

In some European countries, regulators are proposing levels so low that rainwater running off a copper roof would be considered contaminated. Thus copper roofs would be prohibited. By the same reasoning, so would copper pipes. The industry is working with regulators to develop accurate information and, of course, pointing out that copper in the human body is necessary to sustain life. If everything a person drank or ate contained as little copper as is being proposed, sickness and even death would result. I believe the industry will succeed, and that copper will continue in its traditional uses.

Continued demand for the metal in its traditional uses, and its increasing use in the newly expanding economies of places like China and Indonesia will, over the long term, result in a net growth in world consumption of about 2 percent a year. At this rate, a new Grasberg will be required every two years, and P.T. Freeport Indonesia and its competitors in the world of copper will be hard-pressed to fill the demand.

Appendix

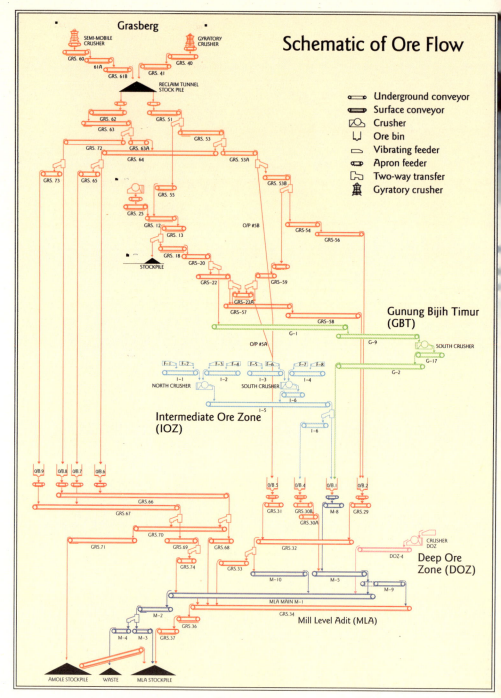

FIGURE A-1

APPENDIX I

The Mine

1. OVERVIEW OF THE ORE DELIVERY SYSTEM

Ore from the Grasberg ore body is crushed to minus 20 cm. (8 in.) in one of the two 152 cm. by 226 cm. (60 in. by 89 in.) Fuller Traylor primary crushers located at the 3,400-meter (11,150 ft.) elevation near the open pit, and conveyed to a 400,000-metric-ton stockpile, also at the mine. Reclaimed ore from the stockpile is then transported to ore passes by either a 183 cm. (72 in.) belt conveying system or a 213 cm. (84 in.) system. The 213 cm. system can transfer 10,000 tonnes per hour; the 183 cm. system, 6,000 tonnes per hour. (See FIGURE A–1 at left.)

Six ore passes deliver ore to the mill at the 2,800-meter (9,200 ft.) elevation. One or two operate at any given time, with the rest on standby. The ore passes are either 2.1 m or 3.0 m (7 ft. or 10 ft.) in diameter, and approximately 600 meters (2,000 ft.) deep. The ore passes perform considerable size reduction on the rock: a typical feed size would be 80% passing 75 mm.; a typical discharge size, 80% passing 44 mm. This feature significantly improves throughput rates in both the crushing plant and SAG mill. (See FIGURE A-3, page 361.)

Ore is reclaimed and directed to either the existing secondary crusher stockpile or to a new SAG concentrator stockpile via reasonably complex, yet fully redundant conveying systems. A number of new tunnels and conveyor ways were required to direct ore onto the new stockpile. Under the SAG concentrator stockpile there are two additional conveying systems; one that continuously reclaims and feeds ore directly to the SAG mill, and another that transfers ore to the secondary crusher stockpile as required. The later system is preferred, as the redundant oreflow/orepass system can be placed on standby, thus reducing wear on the ore pass and associated equipment.

Seven 213 cm. (84 in.) apron feeders reclaim the ore. Three are dedicated to feeding a 183 cm. (72 in.) conveyor which delivers ore to the secondary crusher stockpile. The remaining four feed ore to a series of three 183 cm. (72 in.) conveyors for transfer to the SAG mill feed chute. The belts were designed to transfer 3,320 DMTPH of ore to the SAG mill including recycle tonnage.

2. GRASBERG MINE FACT SHEET, 1996

History
Mine construction began in February 1998
First ore production December 1989

January 1996 10K ore reserve

GRASBERG OPEN PIT
1,152,955,000 tonnes ore

1.06% total Cu	22.1 billion payable lbs. Cu
1.27 g/t Au	34.0 million payable troy oz. Au
2.90 g/t Ag	50.6 million payable troy oz. Ag

GRASBERG UNDERGROUND
604,318,000 tonnes ore

1.20% total Cu	13.1 billion payable lbs. Cu
1.10 g/t Au	15.4 million payable troy oz. Au
3.79 g/t Ag	34.7 million payable troy oz. Ag

Geologic drilling
552 diamond drill holes, totaling 196,500 meters of core
Deepest core hole (GRS 37-246) reaches 1,601.5 meter elevation

Historical Production
In thousand tonnes of ore and overburden

1990	18,059
1991	41,313
1992	59,401
1993	101,174
1994	149,893
1995	189,434
1996 (year to date)	47,311
Total	606,586

1996 Goals
In thousand tonnes of ore

Ore	42,090	1.34% Cu, 1.60 g/t Au
Overburden	159,210	
Total	201,300	

Stripping Ratio 1:3.78 (see FIGURE A-2, page 360)

Safety
Grasberg operations and Grasberg maintenance

Year	LTA	Man-hours	LTA rate	LTA + RCA rate
1990	1	888,505	0.23	0.23
1991	4	1,417,607	0.56	0.99
1992	1	1,826,067	0.11	0.77
1993	6	2,531,551	0.47	0.19
1994	3	2,881,066	0.21	0.35
1995	3	4,073,782	0.15	0.44
1996*	3	1,151,360	0.52	0.69
National Safety Council average			1.40	3.98

*January–April

Personnel
1996 budget

Department	Staff	Non-staff	Total
Operations	71	623	694
Maintenance	61	540	601
Engineering and geology	20	18	38
Total	152	1,181	1,333

Bench Configuration

ORE BENCH

15 meters high, 65 degree face angle, 45 degree overall slope

OVERBURDEN BENCH

15 meters high, 65 degree face angle, 42 degree overall slope

Drilling and Blasting

ORE PATTERN

9m × 8m × 17m
Hole diameter .. 12.75 inches
Explosive ...... ANFO
Powder factor ... 0.15 to 0.25 kg/tonne
Blasthole drill .. D90K

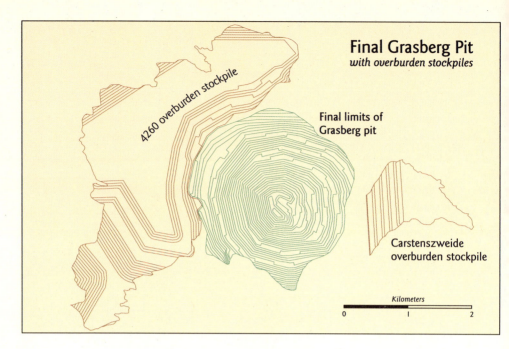

FIGURE A-3

WASTE PATTERN

11m × 11m × 17m
Hole diameter .. 12.75 inches
Explosive ...... ANFO
Powder factor ... 0.12 to 0.15 kg/tonne
Blasthole drill .. D90K

Loading Equipment

Model	Number of units	Capacity	Production (tonnes/hr.)
P&H 4100 shovel	4	42 m3	3,340
P&H 2800 shovel	2	34 m3	3,000
Hitachi EX3500 shovel	3	20 m3	1,390
Caterpillar 992C F.E.L.	3	10 m3	370

Haul Trucks

Model	Number of units	Capacity	Production (tonnes/hr.)
Caterpillar 793	45	218 tonnes	500
Caterpillar 785	21	130 tonnes	300
Caterpillar 777	13	80 tonnes	170

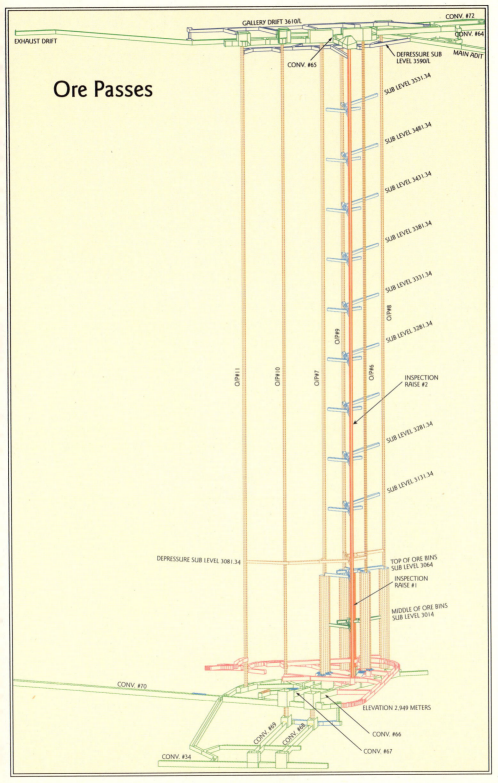

FIGURE A-3

Blasthole Drill and Support Equipment

BLASTHOLE DRILLS

Driltech D90K 5
Driltech D75K 1
Driltech D40 2

TRACK DOZERS

Cat D11N 10
Cat D10N 3
Cat D9N 3

OTHER SUPPORT EQUIPMENT

Cat 992 Tiger rubber tire dozer . . 3
Cat 834 rubber tire dozer 2
Cat 16G grader 12
Lowboy (Cat 777) 1
ANFO truck 5
Explosives service truck 1
50-tonne ANFO trailer 1

2. UNDERGROUND MINE FACT SHEET, 1996

Production

1995, actual   2,358,500 dmt   6,944 tpd   2.04% Cu   0.81 ppm Au
1996, forecast 3,660,000 dmt 10,000 tpd   1.51% Cu   0.52 ppm Au

Manpower
Current

BY NATIONALITY

Expatriate (Including contractors) 51
National staff 81
Hourly non-staff 664
Total 796

BY DEPARTMENT

IOZ operations 402
Maintenance 129
Hilton Shop (service and fabrication) . 93
Capital projects 172
Total 796

Productivity

Direct IOZ .... 149 tons/manshift
Current (total IOZ) .. 22 tons/manshift
By Dec. 1996 .... 57 tons/manshift

Reserves
As of January 1, 1996

Ore body	1,000 tons	%Cu	ppm Au
IOZ	20,886	1.64	0.42
DOZ (Block cave)	49,674	1.72	0.94
Big Gossan	37,349	2.69	1.02
Dom	30,892	1.67	0.42

Development

1995 .... 11,816 meters
1st qtr. 1996 .... 2,308 meters
Total in active use ...> 40,000 meters

Equipment Fleet

LOAD-HAUL-DUMP VEHICLES

Model	Number	Capacity
Eimco 925	10	5 yd.3
Eimco 928	3	8 yd.3
EJC 180	5	5 yd.3
Wagner ST-6	10	6 yd.3

TRUCKS

Model	Number	Capacity
Eimco 985-T15	7	15 tonnes
Wagner MT-420	9	20 tonnes
EJC 430	3	30 tonnes

DRILLS

Model	Number
Ingersoll CMM-2 .	6
Cubex-5200	2
G-D PK-100	1

JUMBOS

Model	Number
MK-65 P	3
MK-65 H	1
MK-45 H	1
MK-35 H	1

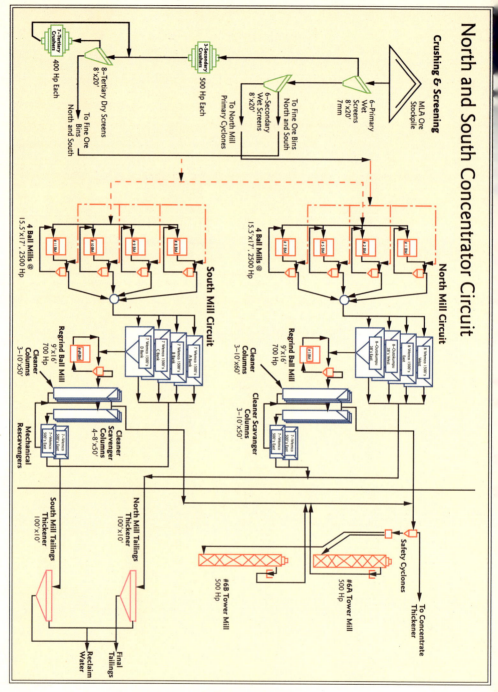

FIGURE A-4

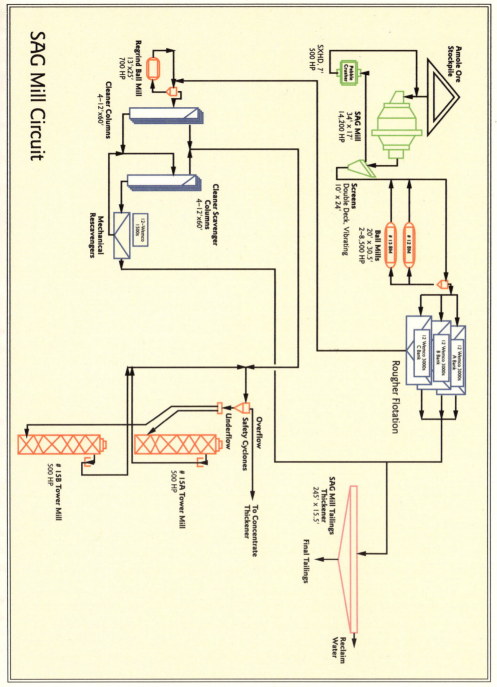

FIGURE A-5

APPENDIX II

The Concentrators

The No. 1 and No. 2 Concentrators

CRUSHING AND SCREENING

Comminution begins in the crushing plant, where the ore is subjected to two crushing stages and three screening stages. The primary wet screens are a set of six Allis Chalmers 7 mm, 8 ft. × 20 ft. screens. The underflow from these screens passes over a set of six Allis Chalmers 4 mm and 2 mm, 8 ft. × 20 ft. screens—the secondary wet screens. The secondary screen underflow constitutes the crusher slurry, which contributes to the North Mill feed (approximately 25% of the dry feed passing over the primary screens reports as crusher slurry). The secondary screen oversize is combined with the tertiary screen undersize in the fine ore bin—the mills' dry feed. The primary screen oversize is fed to the three Symons 7 ft. 500 HP standard cone crushers—the secondary crushers. The secondary crusher product is introduced into the tertiary crushing and screening circuit. This circuit consists of eight each 8 ft. × 20 ft., Allis Chalmers dry screens and seven each 7 ft., 400 HP Symons short head cone crushers (one tertiary screen oversize reports to a secondary crusher). The product is sent to the fine ore bins in which 80% of the ore is less than 8–10 mm.

NORTH AND SOUTH MILLS

With the commissioning of the No. 3 Concentrator, the North and South Mill's tonnage has been optimized, resulting in finer grinds and improved recovery. The facility produces at least a 31% copper concentrate at an 85% recovery with a feed grade of 1.3–1.5% copper.

The North and South Mills are two different entities, although they share many features. Both mills consist of a comminution circuit including four ball mills operating in closed circuit with a dedicated cyclone bank. The flotation circuit includes a set of mechanical roughers producing column/cleaner feed and final tails, followed by cleaner columns, producing final concentrate, a set of cleaner scavenger columns, also producing final concentrate, and a bank of mechanical scavenger cells producing cleaner circuit final tails.

The South Mill draws 28,250 DMTPD from the fine ore bin directly into one of the four primary ball mill's feed chutes. Each is an Allis Chalmers 2,500 HP, 15.5 ft. × 17 ft. mill, charged with 65 mm steel balls. They operate in closed circuit with a dedicated bank of five, 26 in. Krebs hydrocyclones. The cyclone overflow constitutes the flotation feed.

The South Mill's rougher circuit consists of four parallel banks of seven Wemco 1500s with a residence time of 19 minutes. The feed is typically 80% passing 300 microns at a grade of 1.23–1.35% copper. The rougher concentrate is further ground in a 700 HP 9 ft. × 16 ft. Allis Chalmers ball mill, operated in closed circuit with a bank of six 20 in. Krebs hydrocyclones. The reground slurry is fed into three 10 ft. × 50 ft. cleaner columns, operating in parallel, with a mean liquid residence time of 29 minutes. They produce a concentrate of 34–36% copper and a tails of 4–6% copper. The cleaner column tails are combined with the mechanical scavenger concentrate to form a 6–8% copper cleaner-scavenger column feed. The four 8 ft. × 50 ft. cleaner-scavenger columns produce a final concentrate of 24–27% copper with a 4–6% copper tail reporting to two parallel banks of seven Wemco 500 mechanical scavenger cells. The mean liquid residence time within these columns is approximately 9 minutes. The mechanical scavengers' 7-minute residence time results in a 7–9% copper concentrate and a 0.27% tails which reports to final tails. The overall cleaner recovery, from first column feed to mechanical scavenger tails, is 98% with similar recovery around the scavenger circuit.

The cleaner circuit concentrate is further ground in two Allis Chalmers 500 HP tower mills, charged with 13 mm steel balls. These mills are in closed circuit with a bank of six 20 in. Krebs safety cyclones, resulting in a concentrate with 80% passing 50 microns.

The North Mill feed consists of both ore from the fine ore bin and crusher slurry. The total feed is 29,750 DMTPD with crusher slurry comprising approximately 10,000 DMTPD or one third of the feed. The crusher slurry is introduced into the ball mill discharge and thus undergoes classification prior to grinding. Approximately 15% of the crusher slurry will bypass the mills and report directly to the rougher. This results in the same feed rate to the North Mill ball mills as the South's. Crusher slurry is of higher grade than fine ore, resulting in a higher feed grade to the North Mill, approximately

1.35–1.45% copper, and higher recoveries, approximately 85–87%. Each of the North Mill ball mills operate in closed circuit with a dedicated bank of six 26 in. Krebs cyclones. The ball mill product feeds four parallel rougher banks. Two banks of Outokumpu 38s and two banks of Wemco 1500s. The mean residence time within the rougher circuit is 20 minutes with a concentrate grade of 11% copper and a tails of 0.2% copper, at a recovery of 87%. The concentrate is reground in an identical circuit as the South Mill with the product providing feed for three parallel 10 ft. × 60 ft. cleaner columns. The mean liquid residence time within the columns is 47 minutes resulting in a final concentrate grade of 35–38% copper and a tails of 5-6% copper. Similar to the South Mill, cleaner tails are combined with mechanical scavenger concentrate as the cleaner scavenger column feed, approximately 6–8% copper. The three parallel 10 ft. × 50 ft. cleaner scavenger columns have a retention time of 13 minutes, producing a final concentrate of 27–29% copper. The 5% copper tails form the feed to two parallel banks of seven Wemco 500s. The 9-minute residence time within these cells results in a 7–9% concentrate with a 0.2–0.3% tails. The overall cleaner circuit recovery is approximately 98%. The North Mill concentrate joins the South Mill concentrate in the tower mill circuit.

Both Mills use Aerofloat as a primary collector, with sodium isobutyl xanthate (SIBX) as a secondary collector, and methyl isobutyl carbinol (MIBC) as frother. (See TABLE A-1, page 369)

Each mill has a dedicated 100 ft. × 10 ft. Eimco tailings thickener. The tails are 35–40% solids with the thickener underflow at 45–55% solids. The water is reclaimed and used within the grinding circuit.

No. 3 Concentrator

SAG MILL

The No. 3 Concentrator was commissioned in February 1995 increasing site tonnage from 75,000 to 118,000 tonnes per day. It incorporates a SABC—semi-autogenous ball mill crusher—circuit, followed by a flotation circuit with regrind. No. 3 Concentrator has been designed to produce a concentrate of 30% copper or higher at 85–90% recovery from an expected 1996 feed grade of 1.32–1.35% copper.

The No. 3 Concentrator draws run-of-mine (ROM) ore from the Amole Stockpile. The Ore is 80% passing 2 in. and is delivered to the

Allis Mineral System's, 14,200 HP, 34 ft. × 17 ft. SAG mill with a Siemens wrap-around motor, using a combination of apron feeders and conveyers. Lime and 105 mm steel balls are added to this feed automatically. The SAG discharge passes over a 10 ft. × 24 ft. double deck vibrating screen with the oversize reporting to the Pebble Crusher and then back to the SAG mill; the screen undersize flows to the ball mills at 80% passing 1440 microns. The Nordberg 7 ft., 500 HP Short Head, Cone Pebble Crusher sees approximately 14% of the SAG feed and has a closed side gap of 10 mm.

Secondary grinding is accomplished in two Marcy 20 ft. × 30.5 ft., 8,500 HP ball mills operating in parallel. These are manually charged to approximately 35% with 65 mm steel balls. Each mill is operated in closed circuit with a bank of fourteen, 26 in. Krebs fixed apex cyclones. The cyclone feed consists of the SAG's vibrating screen underflow combined with the ball mill discharge. The cyclone overflow, consisting of slurry at 80% passing 150 microns, constitutes the flotation circuit feed with the underflow reporting to the ball mills. Lime and the primary collector are added at this point.

The flotation circuit begins with three parallel banks of twelve Wemco 3000 mechanical flotation cells operating as roughers. They have a residence time of approximately 24 minutes and produce a concentrate of approximately 12.0% copper at a recovery of 94.0%. The concentrate is sent to the regrind circuit where a 13 ft. × 25 ft., 2500 HP Allis Mineral Systems ball mill is run in closed circuit with a bank of ten, 20 in. Krebs cyclones, producing an overflow of 80% passing 70 microns. The mill is charged to 40% with 25 mm steel balls. The overflow from the regrind cyclones constitutes the feed to the cleaner columns. Four each 12 ft. by 60 ft. cleaner columns are run in parallel resulting in a 22-minute mean liquid residence time. They produce a 33% copper concentrate. The 4 cleaner scavenger columns are 12 ft. × 60 ft. producing a concentrate of 23.5% copper at a recovery of 54.5% with a mean liquid residence time of 18 minutes. Both sets of columns utilize the Minnovex Variable Gap Spargers and wash water pans. The cleaner scavenger tails report to a bank of twelve Wemco 1500 mechanical scavenger cells, which have a mean residence time of 18 minutes, producing a concentrate of 9.5% copper at a recovery of 80.13%. The concentrate from these cells is combined with the tails from the cleaner columns to form the cleaner scavenger column feed. The tails are combined with the rougher tails constituting No. 3 Concentrator's final tails. This stream feeds the 245 ft. × 15.5 ft. thickener at approximately 34% solids, resulting in an underflow of 60% solids and a grade of approximately 0.2% copper. The water is reclaimed and used in the grinding circuit.

Sodium isobutyl xanthate (SIBX) and methyl isobutyl carbinol (MIBC) are added to the rougher banks as, respectively, a secondary collector and frother.

The column concentrate must be ground further due to pipeline requirements. The concentrate is ground in a 500 HP, Allis Mineral System's Tower Mill, charged with 13 mm Balls. The mill is run in closed circuit with six 20 in. Krebs cyclones. producing a final concentrate of 80% passing 50 microns.

Thickening, Drying and Dewatering

The thickening, drying and dewatering facilities are located both at the mill and at port site. The concentrate is thickened from approximately 20–25% solids to 65–70%. The North and South Mill's concentrate is processed in two parallel 39 ft. thickeners with the overflow processed in a 39 ft. secondary thickener. The SAG mill's concentrate reports to a 70 ft. diameter thickener. The primary, secondary and SAG thickener underflow feeds six storage tanks from which the concentrate is pumped to port site via 75 miles of 6 in. pipeline. The overflow from the SAG thickener and the North/South secondary thickener is used as reclaim water.

At port site the concentrate is filtered in ten parallel EIMCO rotary disk filters—seven are 8 ft.-10 in. six-disk filters, and the other three are 8 ft.-10 in. eight-disk filters. The filter product is approximately 13–14.5% moisture and is introduced into four parallel dryers. One is a 50 TPH 8 ft. × 40 ft. dryer, the second is 100 TPH, 10 ft. × 60 ft. dryer and the third and fourth are 165 TPH, 10 ft. × 60 ft. dryers. The dryers are fuel/waste-oil fired and produce a final concentrate at 9–10% moisture which is stored in two 50,000 WMT covered barns.

General Reagents Usage
Based on 125,000 tpd throughput

Reagent	Use	gram/ton	Kg/year
Lime	pH modifier	1,900.00	87,000,000
CYTEC S 7249	Collector	22.00	900,000
Aero-3477	Collector	22.00	900,000
SIBX	Collector	7.17	900,000
OTX-140	Frother	24.48	900,000
Magnafloc	Flocculant	10.45	775,000
Deoxer Maxwell	Oxygen scavenger		21,000
Antiscalant	Scale control		180,000

GLOSSARY

Collector
Organic compounds that render selected materials water-repellent by adsorption of molecules or ions on to the mineral surface, reducing the stability of the hydrated layer separating the mineral surface from the air-bubble to such a level that attachment of the particle to the bubble can be made on contact. Hydrophobicity has to be imparted to most minerals in order to float them. Collector reagents—or surfactants—are added to the pulp in the beginning, and some time (a conditioning period) is allowed for adsorption during agitation.

Primary collector
Collector of selective action to float highly hydrophobic minerals. It is usually used at the head of the flotation circuit. Trade names: CYTEC S 7249 or Aero-3477.

Secondary collector
Collector of less selective and more powerful action to promote recovery of slower floating minerals. Trade name: SIBX.

Frother
The name given to a class of reagents that aid in the formation of bubbles or enhance the stability of bubble attachment. These reagents are added when the mineral surfaces have been rendered hydrophobic by the collector reagents. Trade names: MIBC or OTX-140

Flocculant
High molecular weight organic polymers that are used to aggregate suspended particles and cause an efficient separation of solids from the suspending medium. Mineral particles normal settle out of suspension very slowly, but if they can be agglomerated into clumps (flocs) they settle more rapidly. Trade name: Magnafloc

Lime
The name given to a mineral derived from the calcination of limestone followed by hydrating the resulting calcium oxide, or by treating an aqueous calcium salt solution with an alkali. Lime is used to regulate pulp alkalinity prior to flotation, to precipitate heavy metal ions out of solution, to provide stable chemical conditions for the collectors, and to minimize corrosion of the cells and pipework.

TABLE A-1

Acknowlegements

This book was made possible only because a large number of people were so generous with their time and expertise. Among them were:

Les Acton , Richard Alzetta, John Amato, Roger Austin, Stan Batey, Tom Beanal, Wayne Beaupre, Louis Bell, Ali Budiarjo, Yohannes Cenawatme, Dr. Marvin Clark, Wayne Cook, John Cutts, Art d'Aquin, Balfour Darnell, Peter Doyle, Jean Jacques Dozy, Steve Drake, Tom Egan, Fred Elliott, Dave Flint, Del Flint, Dave Francisco, Ron Grossman, Ilyas Hamid, Kris Hefton, Hoediatmo "Dick" Hoed, Eddie Hoffman, Leroy Hollenbeck, Camille Howard, Lawrence Johnson, Mark Johnson, Steve Jones, "Moses" Tembak Kilangin, Nils Kindwall, Gary King, Russell King, Rene Latiolais, R. Douglas Learmont, Howard Lewis, David Lowry, Adrie Machribie, John Macken, John M. Marek, Bruce Marsh, Fred Mason, George McDonald, Jim Bob Moffett, Paul Murphy, Frank Nelson, Gary O'Connor, Jay Pennington, Javier Perez-Fortea, Philip Petrie, Roy A. Pickren, Jr., Kevin Pollard, Dave Potter, Greg Probst, Garland Robinette, Yonaniko Salim, Peter Sedgwick, Dave Singleton, Bud Soehoed, Craig Saparito, Ray Speights, Ted Staines, Barnabas Suebu, Agus Sumole, Wisnu Susetyo, Julius Tahia, Leon Thomas, Hans Tutuarima, Steve Van Nort, Ron Walters, Milton H. Ward, Charlie White, Charlie Wilmot, Kresno Wiyoso, and Lou Zawislak.

Index

References to photographs in italic; references to maps, figures or charts in boldface

acid mine drainage, 271–272
Act of Free Choice, 49
Acton, Les, xvii, 119
adat land rights, 303–304
aeromag, 224; of Wabu, 235
Aetna, 239
Agats, 349
Aghawagong River, 69, 210, 268
Aghawagong Valley, 77, 160, 203
Agrarian Law of 1960, Basic, 303
Agrico Chemical Company, 32
Aika, 62
Aikwa Deposition Area (ADA) 270; map of, **269**
Aikwa River, 268, 290, 322; Colijn expedition on, 67; sheet flow of, 266–267; tailings in, 266–271
Akimuga, 298, 299
Alatief corporation, 182
Allende Gossens, Dr. Salvador, xii
Aluminum, recycling of, 287
Aluminum foil, disposal of, 284
Alzetta, Richard, 231
Amamapare, 171, 295, 352; description of, 179; ore ship being loaded at, 213; scouting original site, 98
AMAX, 30
Amole Adit, 158, 161, 192, 198, 210
Amole Ridge, leveling of, 206
Amungme, 67–69, 290–291, 294, 296–299, 350; agreement by with Freeport, 291–292, 303–304; business ownership among, 312; celebration by of Suharto's 1973 visit, 101; decision-making among, 305–306; education for, 308–310; encounter by with first drill team, 92–94; expectations among, 314; firestarting abilities of, 77; first encounter with by Colijn expedition, 67, 296–297; from Banti 1995, 293; from Nargi in 1936, 292; migration to lowlands of, 298–299; political troubles among, 306; reaction to Western goods of, 299; traditional lifestyle of, 290; woman with child, 301; workers, 322
Anaconda Copper, xi
Andaman Islands, 217
Andina, Chile, xii
ANFO, 192
anhydrite, 191
animal husbandry programs, 314–315
anomalies, geological, 215, 218, 225
Apex Anomaly, 235
apron feeder, 197
Arafura Indah, 338
Arafura Sea, 87, 100, 270, 345
Archbold, Richard, 46
Arjuna, 199
arsenopyrite, 233
Aru Islands, 294
Asia Pacific Services, 182
Asmat, 52, 59; museum of carvings by, 319–320; raids on Mimika by, 295; woodcarving project, 310
Asmat Museum of Art and Progress, 320
Atuka, 295
Australia, tectonics of, 216–218
Australian Plate, 216–218
Austronesians, 34, 40–42
Ayam Hitam Adit, 150

Babo, 61–62
Baitur Rahmin mosque, 334
Bali, 34
Baliem Valley, 46
Balikpapan, 62, 226
Banti Village, 275; Freeport contributions to, 318–319; map of, **309**; pig project at, 318–319; school in, 308; vegetable project at, 316
baramundi fish, 348
Barden, Reg, 87

373

Bataafsche Petroleum Maatschappij (BPM), 60
Bataks, 38
Batavia, 36
Batey, Stan, 314–315
Bea River, 58
Beccari, Odouardo, 46
Bechtel, xiii, 99, 100, 170; first feasibity report from, 102; negative review of pit–steepening plan by, 111
Belakmakema, 299
Bell, Louis, 233, *233*
Beoga, 298
Bergamo Alps, 62
beri-beri, 55
Besi, 70–71
Biak, 290; mine workers from, 195; World War II battle for, 47–48
van der Bie, J.J., 54
Bienemann, Bill, 119
Big Gossan, 160–161, 192, 342; — Fault, 158
Big Men, 306
Bilogai, 227, 235; expansion of base camp at, 238
bird sanctuary, 283–284
black chicken *see* Ayam Hitam Adit
blacksmith shop, 286
Block A, 158, 223; map of geology of, **152–153**
Block caving, 116; diagram of, **125**; techniques of, 200–202
Block I, 239
Boeadi, 278
bondegezou, 278
Borneo, 34, 62
bornite, 115, 207
Borobudur, 35
Bougainville mine, 220
Bougainville, Louis Antoine de, 43
Boundary of 1912, 46
Bowencamp, John, 75
Bratanta, Saleh, 84
Brazos River, 25
Bridge-building, 317–318
British Ornithologist's Union expeditions, 54–59, 296; map showing route of, **53**
Bronze Age, 247
Brooke Hunt, 186

"Bruce's Boys," 293
Bryanmound, 24, 25
Budiarjo, Ali, xvi, xvii, 82, 105
bulldozers, 165
Bulolo River, 219
Bunker, Ellsworth, 81
business incubators, 311–313; building park bench in, 313

cabé, 171
Calcasieu Parish, 24
Call, Rick, 110–111
Caltex, 170
Caminada sulfur mine, 29
Camp II, 322
Campbell, Hamish, 227
Cargo cults, 47, 299–300; wearing of keys in, 300
Carpentry Shop Dam, 210
Carstensz glaciers: map showing shrinking of, **277**; 1936 photograph of, 272; 1995 photograph of, 273; seen from the coast, 44; shrinking of, 274–276; snowbridge of in 1936, 60
Carstensz, Jan, 43, 52, 275, 77
Carstenszweide, 137; as center of mine works, 190; difficulty building drill camp on, 94; first drill team landing in, 91; naming of, 63
Cartstenz Toppen, 295, *see also* Carstensz glaciers
Castro, Fidel, 30
Cemara River, 294
Cenawatme, Yohannes, 311–312
Cenderawasih, University of 315, 322
chalcocite, 159, 207; — cap, 142
Chalcolithic Period, 247
chalcopyrite, 115, 159, 207; in Grasberg ore, 158
Chevron, 61
Chocagoula sulfur mine, 29
Christian and Missionary Alliance, 298
Chuquicamata, Chile, xii
Clark, Dr. Marvin, 307
cleaner columns, 207
climate, monitoring of, 288
clinics, 307
Cochlefelis spatulate, 289
Cock's Comb, *see* Hanekam, Mt.

CODELCO, 352
Coeur d'Alene, Idaho, 244
Colijn, Dr. Anton H., 62–64, 68, 71, 275, 298; at Ngga Pulu, 69
Colijn Expedition, 51, 59, 63–72, 296, 298; at Ngga Pulu, 69; Dayak porters for, 57; establishing base camp, 68; first encounter with Amungme, 67; Kamoro rowers for, 56; map showing route of, **53**; planning air drops for, 66; reaching Carstenszweide, 70; reaching Ngga Pulu, 71–72;
Collinson, Tom, 145
commodity-linked stock, 179
concentrate ships, 179
concentration of ore, 207
concession agreements, 83
The Conquest of Copper Mountain, 75, 78
construction division, 183
contact metasomatic deposit, see skarn
contract of work (COW), 83–86; bilateral nature of, 85–86; original definition of, 83; release of land under, 224; resettlement clause in, 303
conveyors, 196; M-System, 198; SuperSystem, 198
Cook, James, 43
Cook, Wayne, 318
Coore, Bill, 337
copper: anode, 254, 256, 257; blister, 254; boardroom doubts about, 131; cathode, 254; chart of consumption of, 259; chart of price of, 248; consumption of, 260–261; early smelting of, 248–249; electro-refining of, 254; etymology of, 248; history of use of, 247–250; native, 247; price of, see copper prices; recycling of, 256–257; use of in industry, 250;
copper concentrate: composition of, 251–252; contracts for, 258–259; custom market in, 257–258; impurities in, 252; loading of at Amamapare, 249; marketing of, 257–260; treatment and refining charges for, 259–260, **260**
copper prices, 22, 261–262; chart of, **248**; Vietnam-era, xiii
core tray, carrying of to helicopter, 245

Corpus Christi, 26
covellite, 207
Crenshaw, Ben, 337
Crosier mission, 296
Croyn, Vincent "Bob," 75
cryoalgae, 274
Cuajone, Peru, xii
Cuban American Manganese Corporation, 28
Currie, John, 92–94
cuscus, 163; parable about, 232
custom market see copper concentrate, custom market in
Cutts, John, 306, 314, 317
Cyathea tree ferns, 281
Cyprus, early copper works on, 248–249

Dabra, 224
Dalam Diatreme, 156
d'Albertis, Luigi, 46
d'Aquin, Art, 331
Damal, 298, 305; see also Amungme
Dampier, William, 43
Dani, 278, 290, 290
Darnell, Balfour, 87–99
Darnell Ridge, 99, 103
Dayaks, 38, 52; porters for Colijn expedition, 57
Deep Ore Zone (DOZ), 115, 130; holding in reserve of, 198
DeHaviland Dragon aircraft, 62
Demak, kingdom of, 35
Dendrolagus mbaiso, 278
Denison Mines, xii
Development Jumbos, 199
diamond drilling, 230–231; bonus system for, 244; breaking down rods during, 236; rigs, see Longyear drill rigs; rig in Etna bay, 237; rig on Grasberg, 149; trade-off between speed and core, 95
diorite, 155
Dobo, 294, 294
Dom: first drilling of, 159–160; naming of, 100
dongas, 95
Dorei Bay, 45, 46
Douglas, Paul, 30, 102
Doyle, Peter 227–228, 233–236, 233

375

DOZ, *see* Deep Ore Zone
Dozy, Jean Jacques, 21, 51, 59, 62–74, 137, 139, 219, 275, 298; account by of reaching Carstenszweide, 70; discovery by of Ertsberg, 71; map by of Carstensz area, 64; 1995 photograph of, 65; with Colijn expedition 1936, 68; with Colijn expedition at summit in 1936, 69
Drake, Steve, 195
drilling, *see* diamond drilling
dugout canoe, Kamoro, 40
Duke, 182
Dutch East Indies Company (VOC), 36
Dutch in Indonesia, 36
dysentery, 55

Eagle's Nest, 235
East Timor, 38
Easter Island, 35
Eastern Mining *see* P.T. IRJA Eastern Minerals Corporation
Eastridge Discovery Zone, 233, *233*
Ekabu Plateau, 75
Ekagi, 56, 290, 296
El Teniente mine, xi, xviii, 130; 25K expansion of, xii
Ellenberger, John, 305–306
Elliott, Fred, 146, 231, 244–246
Enarotali, 219
Environmental testing, 287–289; indicator animals used in, 288, 289; laboratory used for, 287; of river water, 287; standards for, 289
Ertsberg, 22, 51; discovery of by Dozy, 71; drill camp on, 85; drilling of, 95–96; Japanese courted to develop, 82n.; 1936 photograph of, 73; 1968 photograph of, 88–89; OBM concession on, 72; role of glaciers in revealing, 275; sketch of by J.J. Dozy, 72
Ertsberg East *see* Gunung Bijih Timur
Ertsberg mine: agreement with Wa people on, 291; impact on local population of, 290; plan for underground mining of, 109–110; steepening pit of, 110–111; suggestion that should be shut down, 111
Escondida mine, 222, 258

Etna Bay, 62, 294
Etna Bay prospect, 239–246; base camp of, 239; drill camp in, 243–244; first notice of, 240; fly camp in, 241–243; geology of, 239–240
Etna, Mount, sulfur mines at, 24
exploration geology, 215–246; conditions of in Irian Jaya, 220–221; curtailment of, 127; importance of cross-features to, 223; in Block A, 162–164; Indonesian geologists, 220
extraction drifts, 202
Exxon (Esso), 61

Fairy Lakes Fault, 159
Fakfak, 52
Fanamo village, 75
Faumai Formation, 218
57K Program, 151–154
"Fighting Irish Zone," 228
Filipino miners, 120–121, 125–126, 150, 185, 197–198, 199
Fitzgerald, Ted, alias "Ever Ready," 86–87
Flannery, Timothy, 278
Flint, Delos "Del," 113–114, 142, 159; on Freeport–OBM expedition, 75–78
Flint Extension, 114, 300
Fluor Daniel, 154
Fluor-Utah, xii
Fly River, 39
Foreign Investment Law, 83
Fort du Bus, 44
Francisco, Dave, 168
Frasche process, 24
Frasche, Herman, 24
P.T. Freeport Indonesia (PTFI), 32, 110, 116, 118, 119, 149, 154, 158; call for new reserves, 127, 127n.; concentrate sales of, 258; crisis of profitability, 109; future of, 340–354; privatization program of, 179–182; production of, 186
Freeport Kaolin Company, 30, 31
Freeport Sulphur Company, xi, xii, 23, 26–33; entry into nickel, 28; founding of, 26
Freeport-McMoRan Copper & Gold (FCX), 23, 31, 130, 150, 154, 156,

179, 181; founding of, 130; interest in gold, 131–132; stock offering of PTFI, 132–235
Freeport Minerals Company, xiv, xviii, 30; contract of work with Indonesia, 82–84
Freeport Oil Company, 29, 31
Freeport–OBM expedition, 75–78; canoes of on Mawati River, 77; generating power for radio, 78; reaching Ertsberg, 78
froth flotation, 251; copper concentrated by, 212; technologies of, 207–208
fuel, use of in mine operations, 178

Galveston, 26
gangue, 251
GBT, *see* Gunung Bijih Timur
geofabric, 176
"Geological Results of the Carstensz Expedition 1936," 74
Gibbons, Don, 298
Gillian, Mark, 235
glaciers *see* Carstensz glaciers
gold: boardroom preference for, 131–133; chart of price of, **252**; potential for in Grasberg, 132
Golf tours, 349
Gondwanaland, 215
Goodfellow, Walter, 54–56
Gorong Islands, 294
Grande Ecaille, 26; cutting off pilings for, 28
Grande Isle, 29
Grasberg, 137–164; diagram of formation of, **157**; early analysis of ore from, 139; — Fault, 159; first drilling of, 22–23, 134, 140, 144; first surface samples from, 145; geology of, 155–158; map of faults around, **225**; naming of, 21; receiving news of first cores from, 135; table of drill results from, **148**; vegetation anomaly of, 138; with Carstensz glaciers, 145; with Puncak Jaya, 136
Grasberg mine, 137–164; cutoff grade of ore, 190; first road to, 167–169; infrastructure expansion for, 147–154; ore delivery system for, 150; pit strategy for, 155; pit wall strength of, 191; plan view of **193**; future of, 340–354; shovel in, 141; side view of, **200–201**; stripping ratio of, 190; visibility at, 195
greenbelts, 279
greenhouse effect, 274
Greenvale, 30, 31
Gresik smelter, 254, 258, 340, 349
Grossman, Ron, 84–85
van Gruisen, Jan, 72, 73
Grumman Mallard aircraft, 75, 239, 241
Gunung Bijih Timur (GBT), 300; break to surface of block cave of, 123; closing of, 198; development of LHD operation in, 123; economic importance of, 143; LHD vehicle in, 128; mining of, 115, 118–130; use of slusher system in, 122–123
gyratory crushers, 196

Habbema, D., 52
Habbema, Lake, 46
Hall, Jack, xviii
Hamid, Ilyas, 165, 168–171
Hanekam, Mt., 59, 104
Hanging Wall Fault, 159
Harapan, 322
Haryono, Beni, 231
Hasan Dam, 210
Hatta, Mohammad, 36
Hay Report, 333
Hayes, John, 43
healthy houses *see* rumah sehat
heap leaching, 254–255, 272
Heavy Equipment Access Trail (HEAT), 165–166, 189; aerial view of, 169; building of, 168–173; bulldozers going up, 168; role of in melting glaciers, 274
Hefton, Kris, 162–164
Helicopters: advancing technology of, 79; assembly of in Timuka, 80, 90; geologists working from, 163; high–capacity Russian, 168; hoisting from, *see* hoisting; modified Bell 205, 221; use of at fly camps, 243; use of in Grasberg drilling program, 147

377

Hidden Valley, 181
Highland Valley Copper, 258
Hills, Robert, 82
Himalayan Plateau, 216
Hitalipa, 238
Hoed, Hoediatmo (Dick), xvii
hoisting, 226–227, 228, 229
Hollandia, 47
Hoskins Mound, 25
Hospitals, 306–307
Houston, 26
Huarte, 334
Hughes, Howard, 101
hydroelectric plant, 128
hydrometallurgical nickel smelter, 30
hydromulching, 279–280, 280
hydrothermal alteration, 225

Ibu Besar *see* gyratory crushers
Ice Ages, 275
Idenburg, Mt. *see* Ngga Pilimsit
Ilhas dos Papuas, 43
Inauga River, 324
Indonesia: colonial period of, 36; communist party of (PKI), 38, 81; history of, 33–38; land ownership in, 303; map of, **33**; migration of people into, 34–35; religions of, 38
Intermediate Ore Zone (IOZ), 115, 130, 198–199; rock breaker at work in, 124
intrusion, formation of, 219
intrusive float, 226
Irian Barat, 79, 81
Irian Jaya, 38–50; agriculture of, 42; colonial government of, 52; early European settlement of, 43–44; exploration of, 46; exploration of, map, 53; flora and fauna of, 278–279; human migration to, 40; increase of population of, 290; indigenous people of, *see* Amungme, Kamoro; Indonesian claims to, 48; Indonesian development of, 49–50; Indonesian paratroopers in, 80; Japanese soldiers in, 47; languages of, 39; map of, **41**; naming of, 106; transfer to Indonesia of, 79–82; World War II in, 47
Irian–Papuan Fault Belt, 158

P.T. IRJA Eastern Minerals Corporation, 223–224, 238–239
Iron Age, 249
Iwaka, 319

Jakarta, 33
Jansz, Willem, 43
Jaramaya River, 175
Java, 34
Javanese business etiquette, 83
Jayapura, 347
Jefman Island, 61
Jerritt Canyon gold mine, 31, 32
Julius, 293–294
junction sampling, 226
"Jurassic Park," 239
Jurassic Period, 217

Kais Formation, 218
Kaiser Wilhelmsland, 52
Kali Dikes, 156
Kalimantan, 34
P.T. Kamanta, 337
Kamerer, Father Michael, 298–299
Kamora Jaya, 326
Kamoro, 279, 290–291, 294–296, 324; business ownership among, 312; longhouse, 297; modern houses for, 305; traditional lifestyle of, 290; workers in environmental department, 283
Kapare River, 55
Karaka, Pulau, 319
Karang Senang, 326
Karapao, 297
Karubaga, 235
Kemabu, 238
Kembalangan Formation, 218, 224
Kembu River, 224
Kennecott, xi
Kidder Peabody, 135
Kilangin, "Moses" Tembak, 76, 298–299, 322; negotiating for Freeport with Amungme, 92–94
Kindwall, Nils, 31, 84, 102
King, Gary, 196
king-of-tides, 87
Kiruru (Etna) Village, 239
Klamono anticline, 61
Kloss, C. Boden, 58

Kokonau, 171, 295, 298
Kolff, Dirk Hendrik, 44
Koot, J.P., 76
Koperapoka Baru, 324
Kopi River, 266
Koranga Creek, 219
Korean coal miners, 105
koteka, see penis gourd
Krakatoa, 34
Kreditanstalt für Wiederaufbau, 102
Kroesen, A.J., 295
Kuala Kencana, 279, 313, 330–339, 345; *alun-alun* of, 334, 336; building of, 330–339; concept, 330; inauguration of, 331, 332; map of, **333**
Kucing Liar, 342
Kwamki Baru, 324
Kwamki Lama, 308, 311, 324; lakes near, 283; market of, 320; settlement of, 322–323

La Freniere, Don, 90–93, 98–99
Lake Wilson, *133*, 191, 210
Laloki River, 219
landslides, 178
larviciding, 307
Laurasia, 215
Learmont, Doug, 310
leeches, 241
Leiden Journal of Geology (Leidsche Geologische Mededeelingen), 21
Lembah Tembaga, 161, 342
Lesser Sundas, 34
Lewis, Howard, 283
Light Industrial Park (LIP), 331, 338, 348
"Limestone Cowboys," 228
littering, 284
Little Ice Age, 275
load-haul-dump (LHD) equipment, 202
Lokeito, 126
London Metal Exchange (LME), xiii
Long, Augustus C., 100
Longyear diamond drill rigs, 94, 114, 146, 230; use of LF70, 244–246
Long-hole Jumbos, 199
Long-term Environmental Monitoring Program (LTEMP), 288–289
Lorentz, H.A., 52
Lorentz Reserve, 270, 345, 349
Lossman, Horst, 176

MacArthur, Douglas, 47
Macken Dam, 210
Macken, John, 183
Mackenzie, Doug, 227
Maclay, Nikolai Miklouho-, 46
Macrobranchium shrimp, 289
Magal, Paulus, 322
Main Grasberg Intrusion (MGI), 156
Main Pass Block 299, 32
Majapahit Empire, 35
Malacca, 36
Malaria, 55, 306; program to control, 307–308
Maluku, 34
Manokwari, 46, 52
Mapurujaya, 322
Marco Polo, 35
Marek, John, 143
Marind, 59
Marsh, Bruce, 267, 293–294
Marshall Fund, 36
Marshall, E., 57
Mason, Fred, 190
massoi bark, 294
matte, 254
Mawati River, 76
mbaiso, 278
McDonald, George, 155–156
McMoRan Oil and Gas, 31
McNally, Dick, 101
medical waste, disposal of, 287
Meervlakte, 218
Meneses, Jorge de, 42
Merauke, 52, 338
Meren glacier, 276
Merkusoord, 44
Mesozoic Era, 215
Mile 21, 282
Mile 23, 282
Mile 32, 330
Mile 50, 322, 330
Mile 57, 177
Mile 71, *see* Ridgecamp
Mile 74, *see* Mill
Mill: additions to for 57K Program, 151; ball mills at south concentrator of, 205; computerized monitoring of, 206–207; creating space at, 167; tramp material in, 203; diagrams of concentrate flow, **364–365**;

379

Mill (*continued*): leveling ridge to enlarge, 173–174; map of **208–209**; maximizing production of, 204; operation of, 203–204; reagents used at, table, **371**; SAG, *see* semiautogenous grinding mill; technical data on, 364–371; water shortage at, 210; training, program at, 211; upgrading for Grasberg ore, 149

Mill Level Adit, 126, 128, 210

Mimika, 294–296, 324; dispirited nature of, 296; Dutch colonization of, 295; historic trade of, 294; woodcarvings of, 294; World War II impact on, 296; *see also* Kamoro

Mimika River, 54, 55; exploring, 295

Minajerwi River, 266–271

Minart, Larry, 234

mine, *see* Ertsberg mine, Grasberg mine

miners: Filipino, *see* Filipino miners; Irianese in underground mine, 112; Korean coal —, 105; number at Grasberg, 190; running shovel, 194–195; troubles with North American, 120

Mingangkabau, 170

Mining Engineering, xi

Mining, early technologies of, 251

Minjauh, 238

Misool Island, 61

Missionaries, 38, 46–47

Mitsubishi, 154, 253

MLA, *see* Mill Level Adit

Moa Bay Mining Company, 29; expropriation of, 30

Mobil Oil, 61

Moffett, James Robert "Jim Bob," xvii, 22, 31, 135, 181; and trendology model, 222 ; call for reserves by, 127; new philosophy brought by, 127; supports exploration 236, 238; with Suharto, 332

Moni, 278, 290

mud crabs, 348

Mukumuga River, 268

Murray, Joe, 119

muskeg, 91

Nabire, bald hills near, 138

Namo, 293–294

Nargi hamlet, 67, 92; Amungme men from, 292

Narkime, Tuarek, 94

National Potash Company, 29

Nawaripi Village, 305, 319–320

Naziarta, Hazan, 194–195

Nederlands Nieuw Guinea Petroleum Maatshcappij (NNGPM), 60–61

Nelson, Frank, 114–115, 139, 142, 159, 300

New Albion, 43

New Guinea: colonial borders of, 44–45, 52; European contact with, 42; naming of, 43; pacification of people of, 59; tectonics of, 216–218; trade products of, 42, 43; *see also* Irian Jaya

New Town, *see* Kuala Kencana

New Wa, 304

New Zealand Pass Fault, 159

Newmont, xi, 219

Ngga Pilimsit, 276

Ngga Pulu, 59, 275; summit of reached by Colijn expedition, 71

Nicaro Nickel, 28

No. 3 Concentrator *see* mill *and* semiautogenous grinding mill

Noord River, 52

Nord-deutsche Affinerie, 102

North Concentrator *see* mill

Northwall Firn, 275

Nosola River, 54, 58, 300, 302

van Nouhuys, J.W., 52

Numa Valley, 298–299

O'Connor, Gary, 227, 232, 235, 240, 243

oil strike, 61

oil, waste, 286

Ok Tedi mine, 134, 220, 222, 258

Omawita village, 75

Omawka River, 75

Omba River, 239

115K Program, 165–185

141st parallel, 44

Oost Borneo Maatschappi (OBM), 72

OPEC, 34

Opitawak hamlet, 67, 298

Orden Baru, 38

ore: economic definition of, 142; reference to at Etna, 240

380

ore flow system, 196–197; diagram of **356**; for Grasberg, 174–175
ore passes, 129–130, 150, 197; blockage in, 197; diagram of, **361**
ore recovery, rate of, 211
ore reserves, table of, **162**
ore stockpile, 184
Organisasi Papua Merdeka (OPM), 50, 306; cutting of slurry pipeline by, 214, 306; military actions in retaliation against, 306
Otakwa River, 54, 58
Otomona River, 268, 287, 317
outrigger canoe, invention of, 35
overburden: buffering capacity of, 272–273; diagram of dump locations, **360**; management of, 271–273

P & H, 192; *see also* shovels
Pacific Plate, 216–218
Pad 11, 175
Paiva hamlet, 67
Pakpahan, Rachmat, 203, 211
Pamuntjak, Usman, xviii
pandanus, 279
Pangaea, 215
Paniai Lakes, 46, 75, 290; first naming as Wissel Lakes, 63
Papua New Guinea, 39; gold rush, 219
Papuans, 21
parang, 302
Parimau Village, 55, 56
Park, William "Shark-eye," 219
Pasar Raya, 327
Pasar Swadaya Murni, 327
PBY seaplane, 100–101
penis gourd, 21, 39, 77
Perkand Elmer, 287
Peterson, H.I. "Pete," 87
Petrie, Philip, 224, 241–243
Phelps Dodge, xi
Philippines, 35
Phragmites grass, 276, 282–283; use of in shaping rice paddies, 283; use of in stabilizing tailings, 282
pilings, cutting by hand, 27, 28
Pincock, Allen, & Holt, 110
Plain of Lakes *see* Meervlakte
Plate Tectonics, 215–218; map of around New Guinea, **217**
Pliocene Epoch, 220
plumbing system, faults as, 219
poker chip rock, 191
Pollard, Kevin, 182
Pomako, 317
Pomeroy, J.H. company, 99, 100; first feasibity report from, 102
Popular Unity Party, xii
porphyry: characteristics of in different parts of the world, 222; history of, 155; magnetic characteristics of, 225; geological nature of, 113; Pacific model of, 141–142
Port Moresby, 39, 45, 219
Port Sulphur, 27, 29
porters, early explorers' problems with, 55
Portuguese, 36
Potter, Dave, 22, 134, 168, 220, 224, 228; early work on Grasberg by, 137–145
Poydras Street, 31
Prambanan, 35
privatization, 180–183
production-sharing agreements, 83
protore, 137, 142
Puncak Jaya, 39, 51, 275, 295

quartzite, 138

Rawling, Cecil G., 55–57
reagents, 207; chart of used at mill, **369**
Recycling, 284–287; of copper, 256; of old tires, 285; of waste oil, 286
de Retes, Yñigo Ortiz, 43
rice, growing of in tailings, 282
Ridgecamp, 177
Rimba Irian Golf Course, 337
Rio Tinto Minera (RTM) smelter, 253
risk, managing, 183
Road Header, 199
road, mine access: aerial of, 348; amount of use of, 178; building, 103–105; extending to port, 175–176; maintenance of, 177–178; reinforcement of with old tires, 285; scouting original route, 98–99; sinking of during building, 103;

road, mine access (*continued*): strengthening, 176–177; with glaciers in background, 49
road, trans-Irian, 346; Nabire–Ilaga sector, 346
rock art, 37, 40
Rockefeller, John D., 24
Rock-breakers, 202
Royal Dutch/Shell, 60
Royal Netherlands Geographical Society, 295
rubber growing, 295
Rukun Tetangga, 338
Rukun Wilayah, 338
rumah sehat, 304, 319
Rumaropen, Ontel, 195
Russell, Bob, xvii
Ruygrok, Jan, 75; generating power for radio, 76

salib see taboo sticks
salt domes, 24, 25, 29
salt-and-pepper organization, 154, 154n.
San Manuel mine, 119
Sansaporese porters, 76
Santos, 197
satellite images, 223
Satuan Pemukiman, 326
Satwa Indah development, 337, 338
schools, 308–310, 312; vocational, 310
Schouten, Ir. C., 74, 139
sediment, in Irian Jaya's rivers, 267
semiautogenous grinding mill (SAG), 174, 204; diagram of flow through, **365**; operation of, 206; technical data on, 368
Sempan, 290, 294, 324–325
Sempan Barat, 324
Seram Island, 294
Seremoek, 76
shanty town, 304
sheet flow *see* Aikwa River, sheet flow of
Sheraton Inn Timika, 283
Shotcrete, 202
shovels: at Grasberg, 192; hauling of to mine, 165; in Grasberg pit, 188; models working at mine, 173; P&H 4100, working in, 194–195; welding bucket of, 172

Sikorsky S–38 aircraft, 61, 62
silver, chart of price of, **253**
Simpang River, 67
skarn: capricious nature of, 113 ; formation of at Ertsberg, 112; structure of at Wabu, 234; why none at Grasberg, 155–156
slurry, contents of, 212
slurry pipelines, 212–214, 349; building of, 106; cutting of by OPM, 214, 306; tuning of, 213; wear of, 213
Soehoed, Bud, 334
Solomon Microplate, 218
solvent extraction–electrowinning (SX–EX) *see* heap leaching
Sorong, oil deposits near, 219
South Concentrator *see* Mill
South Grasberg prospect, 161
South Kali drill hole, 144
Southern Clays, Inc., 30
spice trade, 36
spilling, 150
Srikandri *see* Road Header, 199
Sriwijaya Empire, 35
Standard Oil, 24
Stewart, Bob, 139–141; early report on Grasberg by, 141
Stewart, Dick, 120, 126
stock offering, abortive London, 132
Stone Age, 247
stream sampling, in Etna, 242–243
subduction, 217
Sudirman Range, 274, 321
Sugapa, 227
Suharto, 34, 38, 49; becomes president, 82; naming of Irian Jaya by, 106; visit to Irian by, 331; with Ali Budiarjo, 105; with J. B. Moffett, 332; with Forbes Wilson, 104
Sukarno, 36–38, 48; fall from power of, 81–82; seizure of Dutch assets by, 80, 81
Sulawesi, 34
sulfur: delivery of liquid, 29, history of use, 23–26
Sumatra, 34
Sumole, Agus, 315–317
supplies, volume of used in mine operations, 178
surge pile, 196

survey pegs, theft of by Irianese, 232
Susetyo, Wisnu, 265, 287–289
sustainable development, 314–318
Sutowo, Ibnu, 82
Suyatno, Pardal, 243–244
swamp forest, lowland, 45
Swenson, Eric, 25, 26

taboo sticks, 91–94, 300–302; negotiating for removal of, 92–94; placed around Ertsberg, 91–94
Tadong, 70–71
Tahija, Julius, 82
tailings, 265–271; farming of, 281–283; making brick from, 264; plan for management of, 268–271; resource created by, 271; thickener, 173, 211
Taiwanese, aboriginal, 35
Tanah Merah, 59
Tapiro Pygmies, 56, 296
Tariku River, 224
Tarpon Lodge, 26
Tembagapura, 49, 181, 290, 302, 345; attractiveness to highlanders, 304; limitations of for housing workers, 327; map of, 180; naming of, 106; scouting original site for, 98
Tethys Sea, 215
Texaco, 61, 82
Texas City, Texas, 154
32K Program, 147–150
Thomas, Leon, 197
Timika, 49, 290, 345, 347; history of, 321–330; map of, **328–329**, ; migrants to, 323; origin of, 323; shopkeers at market of, 324
Timika Indah, 325
Timika Jaya, 326
Timika Pantai *see* Timuka
Timini, 241
Timuka village, 81, 86, 321
Tipuka River, 62, 86, 322
Tirta Indah, 338
Tom, 227
Toraja, 38
tourism, 283–284
training, program of at mill, 211
tramp material, 203
tramway: construction of bases for, 93; foundations of being built, 97; lightening ore buckets of, 124; ore bucket on, 108; severe oscillations of, 105–106
transmigration program, 49, 326
tree kangaroo, 278
trendology model, 222
Trenkenschuh, Father Frank, 296
treponema, 306
Trikora, Gunung, 52–53, 270, 275
Triton Bay, 44–45
trucks: Caterpillar 793B, 196; heavy Caterpillar, 161, 170, 171; running, 195; working at Grasberg, 192
Tsing River, 54, 302
Tsing Valley, 298–299
Tsinga, 298–299, 302
Tsolme, Paulus Zaelingki, 299
Turtle, LCT boat, 86
20K Project, 128–131

Uncle Sam, Louisiana, 32
Union Sulphur Company, 24
United Nations, 49
University of Cenderawasih, 347; scholarship program to, 310
Upper Wa, 304
d'Urville, Dumont, 43
Utekini hamlet, 77

valve, water, plated with copper, 144
Vaughan, Thomas, 30
vegetables, consumption of, 178
Vietnam War, 79

Wa Valley, 229, 290–291, 298, 302
Wabu prospect, 233–239; drilling at, 230; first identification of, 235; geological nature of, 234; helicopter and drill pads on, 216
Wakatimi Village, 55
Wallace, Alfred Russel, 45
Walter, Ron, 212–214
Wanagong Fault, 159
Wanagong Lake, 143
Ward, Milt, xiv, xvi, 31, 110, 126, 138, 146
Waripi Formation, 218
water, use of at mill, 210–211
water quality, standards of, 287–288
van de Water, R., 58

Van de Water Glacier, 59
Wemanawe, 295
West Hanging Valley, 114
Western Otomona River, 67
White, Charlie, 292, 306
Whitman, Rod, 337
Wilhelmina, Mt. *see* Trikora, Gunung
Williams Company, 32
Williams, Jr., Langbourne, 26, 82
Williams, Tommy, xvii, 168
Wilmot, Charlie, 174, 203–205
Wilson, Alex W., 274
Wilson, Forbes K., xvii, 22, 30, 51, 82, 84, 95, 99–101, 299, 302; on Freeport–OBM expedition, 75–78; reads Dozy's report, 74; with Suharto, 104
Wintraeken, Augustinius "Gus," 75
Wisnumurti Range, 276
Wissel, Frits J., 63–64, 68, 71; at summit of Ngga Pulu, 69
Wissel Lakes, *see* Paniai Lakes
Wiyoso, Kresno, 311
Wollaston, Alexander Frederik R., 54–59
Wollaston Glacier, 59
work force, 185; number in at mill, 211
World Health Organization, 287
World War II: in Indonesia, 36; in Irian Jaya, 47–48; old gun emplacement from, 48; Yabu, 238

Yamakupu, 62
Yapen Island, 61
yaws *see* treponema
Yayasan Irian Jaya, 310
Yellow Valley, 71
Yellow Valley Axial Fault, 159

Zaagkam, Mount, 104